Perry L. Miller

Expert Critiquing Systems
Practice-Based Medical Consultation
by Computer

With 33 Illustrations

Springer-Verlag
New York Berlin Heidelberg Tokyo

Perry L. Miller, M.D., Ph.D.
Department of Anesthesiology
Yale University School of Medicine
New Haven, Connecticut 06510
U.S.A.

Series Editor
Bruce I. Blum
Applied Physics Laboratory
The Johns Hopkins University
Laurel, Maryland 20707
U.S.A.

Library of Congress Cataloging in Publication Data
Miller, Perry L.
  Expert Critiquing Systems.
  (Computers and Medicine)
  Bibliography: p.
  1. Medicine—Decision making—Data processing.
2. Medical logic—Data processing. 3. Expert systems
(Computer science) I. Title. II. Series: Computers
and medicine (New York, N.Y.) [DNLM: 1. Computers.
2. Patient Care Planning. W 26.5 M649e]
R858.M543 1986   610'.28'5   85-31760

Typeset by Publishers Service, Bozeman, Montana.

9 8 7 6 5 4 3 2 1

ISBN-13:978-1-4613-8639-1   e-ISBN-13:978-1-4613-8637-7
DOI: 10.1007/978-1-4613-8637-7

Series Preface

Computer technology has impacted the practice of medicine in dramatic ways. Imaging techniques provide noninvasive tools which alter the diagnostic process. Sophisticated monitoring equipment presents new levels of detail for both patient management and research. In most of these technology applications, the computer is embedded in the device; its presence is transparent to the user.

There is also a growing number of applications in which the health care provider directly interacts with a computer. In many cases, these applications are limited to administrative functions, e.g., office practice management, location of hospital patients, appointments, and scheduling. Nevertheless, there also are instances of patient care functions such as results reporting, decision support, surveillance, and reminders.

This series, Computers and Medicine, will focus upon the direct use of information systems as it relates to the medical community. After twenty-five years of experimentation and experience, there are many tested applications which can be implemented economically using the current generation of computers. Moreover, the falling cost of computers suggests that there will be even more extensive use in the near future. Yet there is a gap between current practice and the state-of-the-art.

This lag in the diffusion of technology results from a combination of two factors. First, there are few sources designed to assist practitioners in learning what the new technology can do. Secondly, because the potential is not widely understood, there is a limited marketplace for some of the more advanced applications; this, in turn, limits commercial interest in the development of new products.

In the next decade, one can expect the field of medical information science to establish a better understanding of the role of computers in medicine. Furthermore, those entering the health care professions already will have had some formal training in computer science. For the near term, however, there is a clear need for books designed to illustrate how computers can assist in the practice of medicine. For without these collections, it will be very difficult for the practitioner to learn about a technology which certainly will alter his or her approach to medicine.

And that is the purpose of this series: the presentation of readings about the interaction of computers and medicine. The primary objectives are to describe the current state-of-the-art and to orient medical and health professionals and students with little or no experience with computer applications. We hope that this series will help in the rational transfer of computer technology to medical care.

Laurel, Maryland   BRUCE BLUM

Acknowledgments

This research was supported in part by NIH Grants LM03798 and R01 LM04336 from the National Library of Medicine, and was performed in part using the Rutgers AIM computer of the SUMEX computer network, supported by the NIH Division of Research Resources.

The author would like to thank a number of people who have helped in the projects described in this book. Particular acknowledgment is due Dr. Henry R. Black, who served as the domain expert for both the HT-ATTENDING and PHEO-ATTENDING systems, and to Steven J. Blumenfrucht, a Yale medical student, who implemented PHEO-ATTENDING.

The author would also like to thank Ms. Laurie Hauer for her help in preparing the initial manuscript.

Chapters 4, 5, 6, and 7 were adapted with permission from articles written by the author (P.L. Miller and Black 1984; P.L. Miller 1985a; P.L. Miller et al. 1984; and P.L. Miller 1985c respectively).

Contents

Chapter 1

Expert Critiquing Systems

Over the past 25 years, computers have brought major changes to medicine. These changes are most evident in the administrative side of medical practice. The use of computers has become well established in hospital financial record keeping, and also in more patient-oriented administrative functions such as admitting, bed allocation, etc. In addition, the computer is being used increasingly to store clinical data, for instance, for laboratory test reporting, and in pathology and radiology information systems.

Throughout this evolution, there has been a sense by many workers in the field that the most far-reaching and exciting use of computers would be in helping the physician with the clinical decisions of patient care. The computer has a great deal to offer the clinician. First of all, the computer can accurately recall vast quantities of detailed information. Since the amount of knowledge relevant even to a subspecialty area of medicine is more than most humans can retain, the computer has clear value, provided it can convey its knowledge to the physician effectively.

In addition, medical knowledge is in constant flux as new drugs are developed, new studies are performed, and new papers are written. The computer has the potential to make this information available in an up-to-date form to practitioners anywhere in the world.

Indeed, over the past 20 years, a large number of research groups have explored how computers might be designed to assist the physican in both diagnosis and treatment. This task has not proved to be easy. Medical knowledge can be very complex. It is a hybrid of underlying scientific concepts, clinical hypotheses based on experimental data, and anecdotal experience compiled over years of clinical practice. This diverse knowledge does not fit easily into a simple, clean framework.

Another problem is the great latitude for variation and subjective judgment in medical practice. This is particularly true in areas of treatment. Although there are usually clearly "wrong" ways to treat a patient, there is seldom a single treatment that is absolutely and unequivocally the "right" choice.

Granting the complexity of underlying medical knowledge and of medical practice itself, it is important to define how the computer can best help the physician deal with this complexity. The research described in this book addresses this question. The book describes the exploration of a particular approach to bringing computer-based advice to the practicing physician: a *critiquing* approach.

Although the clinical utility of critiquing has yet to be assessed objectively, it is our conviction that the approach offers computer-based advice in the form most useful for the practice of medicine, and indeed for many other areas beyond medicine where decisions involve a significant amount of subjective judgment.

1.1. Exploring the Critiquing Approach

This book describes four expert computer systems designed using artificial intelligence (AI) techniques to *critique* a physician's plan for patient care. Compared to previous AI systems in medicine, critiquing represents a different approach to bringing computer advice to the practicing physician.

Previous expert systems have attempted to *simulate* a physician's decision-making behavior. They accept as input information about a patient, and produce a set of conclusions and recommendations. Depending on the medical domain, this output may be a list of possible diseases, a set of suggested tests to perform, or a proposed treatment regimen. Such a system has the clinical effect of trying to tell a physician what to do.

As outlined in Figure 1.1, a critiquing system differs from this traditional approach in that it accepts as input both information describing a patient and a physician's proposed plan. The system then critiques the plan. In this way, it structures its advice around the physician's own thinking and style of practice.

The research described in this book explores how critiquing may be performed, and how a computer system is best designed to critique medical practice effectively. The method chosen to explore critiquing may seem unusual to readers not familiar with computer science research. One approach might have been to con-

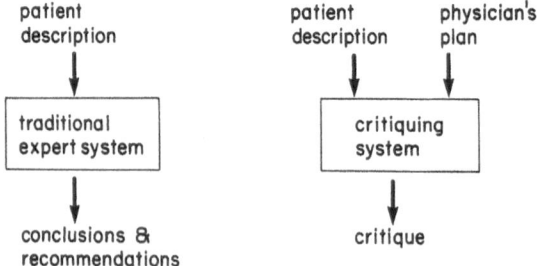

Figure 1.1. A critiquing system bases its analysis on both (1) information about a patient, and (2) a physician's plan.

centrate on single domain and develop a practical consultation tool, before going on to explore critiquing in further domains.

We chose not to focus in depth on one domain. Instead we have implemented several prototype systems in quite different areas of medicine. As discussed further in Section 1.4, this strategy has afforded a broader, more comprehensive understanding of the critiquing process.

The book describes four prototype systems and the lessons learned from their implementation. The remainder of this introductory chapter gives an overview of this work, to help put the different systems into an overall perspective. Chapter 2 then discusses the field of artificial intelligence in medicine as a whole. Chapters 3 to 10 discuss the individual systems and their implications in more detail.

1.1.1. ATTENDING (Anesthesiology)

The first system to implement the critiquing approach was ATTENDING, whose domain is anesthetic management. Central to ATTENDING's design is *risk*. Anesthesiology is a high-risk specialty. Any medical problem that a patient has may imply risks for that patient's anesthetic management, frequently life-threatening risks. A particular anesthetic technique may increase a given risk or may decrease it. In complicated patients with several medical problems, there are frequently risk tradeoffs, where techniques good for one problem are bad for another, and vice versa.

As a result, in planning anesthetic management, a physician must weigh these various risks against one another. In so doing, he tailors his anesthetic management to the patient's underlying problems and to the risks that they imply. ATTENDING is designed to assist in this process.

Because anesthetic risks are so immediate and so critical, anesthesia is an excellent initial domain for exploring the critiquing approach. The anesthesia domain has served to focus attention upon risk, and on the process of risk analysis, as a central underlying paradigm of medical management. Indeed, as described in Chapter 3, a heuristic approach to risk analysis was developed to allow ATTENDING to deal explicitly with risk in its internal analysis.

Although anesthesia is a good initial domain for exploring the critiquing approach, it has significant drawbacks as a domain for implementing a practical consultation system. The major problem is the size of the domain: the large amount of knowledge required to create a truly "expert" consultation system. The knowledge from several volumes on anesthesia, on anesthetic agents, and on the anesthetic implications of many different diseases would be required.

The problems involved in collecting all this knowledge, incorporating it into the system, and validating its accuracy would be large. Since both the optimal design and the best advice-giving features for a critiquing system are still being explored, it is premature to tackle such a large volume of knowledge, which would later require iterative reworking as fundamental design changes were made. It is much better to crystallize solutions to the various design issues in smaller domains.

As a result, instead of pursuing the domain of anesthesiology, we decided to concentrate on smaller, more constrained domains where a comprehensive expertise might more easily be achieved. In exploring these domains, an added benefit was that further interesting facets of the critiquing approach were uncovered.

1.1.2. HT-ATTENDING (Essential Hypertension)

The second domain chosen to explore critiquing was the pharmacologic management of essential hypertension. Here, the HT-ATTENDING system critiques the various drugs and combinations of drugs proposed for the outpatient treatment of a patient with high blood pressure.

Hypertension is an interesting domain from a number of standpoints:

1. It is a widespread problem that the practicing physician encounters frequently.
2. There is a truly bewildering armamentarium of drugs that can be used. These include diuretics, beta blockers, central adrenergic inhibitors, peripheral sympathetic blockers, vasodilators, angiotensin converting enzyme inhibitors, and calcium channel blockers. Indeed, there are several drugs in each of these categories, each with its own indications and contraindications in the presence of various concurrent medical problems.
3. Finally, the field of hypertension management is in rapid flux. New drugs and new types of drugs frequently appear, and new studies are performed evaluating the efficacy of various drugs and combinations of drugs in different classes of patients.

In the face of this onslaught of new knowledge, it is virtually impossible for the practicing physician to keep completely current in the field, especially since the hypertensive patient represents only a fraction of his practice.

A physician deals with the modern information overload in part by basing his practice on a subset of medical knowledge. He develops an approach that he is comfortable with: where he is familiar with the major indications and contraindications for the different agents he uses, and where he therefore feels that he can treat his patients safely. For the large majority of cases, such a restricted style of practice may work well. It is nevertheless useful for the physician to get periodic feedback and exposure to new agents that he may like to incorporate into his practice.

HT-ATTENDING is designed to give this feedback by discussing how a physician's style of hypertension management fits in with current thinking in the field, and with the current spectrum of agents available for treatment. Thus HT-ATTENDING plays two roles. It provides feedback concerning the physician's approach to a particular patient. It also provides a form of practice-based education structured around daily patient care.

HT-ATTENDING can be envisioned as an "interactive paper," a computer-based paper that structures its content around a particular plan to a patient's care. Instead of skimming a review paper on hypertension management for sections relevant to a particular patient, a physician can use HT-ATTENDING to do this for him, and to structure the relevant material into a focused analysis.

HT-ATTENDING's domain has the further advantage (compared, say, to ATTENDING's domain of anesthesiology) that it is reasonably constrained in its scope. Whereas ATTENDING's domain would require several volumes of material to be complete, the outpatient management of hypertension requires roughly the contents of one or two survey articles. As a result, we anticipate that sufficiently comprehensive expertise can be incorporated to let HT-ATTENDING function as a practical consultant. This is a major current goal.

Another interesting feature of the hypertension domain is the rapid flux of clinical knowledge. In the 2 years since the initial HT-ATTENDING prototype was implemented, recommended treatment has changed in several significant ways, and several new drugs are available. As a result, the system has been updated to reflect the current state of the field and will have to undergo similar revisions in the future.

1.1.3. VQ-ATTENDING (Ventilator Management)

The VQ-ATTENDING system explores a third area of medical management: the ventilator management of a patient receiving mechanical respiratory support. The rationale for exploring several domains of medical management (anesthesia, essential hypertension, and ventilator management) was partly to demonstrate that the critiquing approach was generally applicable, and not useful only, say, in anesthesiology. The domain of ventilator management is particularly interesting, however, because it points up a style of critiquing design that has not been readily apparent in the previous domains: "goal-directed" critiquing.

In designing VQ-ATTENDING, it was found very helpful to separate two types of knowledge about the domain: (1) *strategic* knowledge about the treatment goals of ventilator management, and (2) *tactical* knowledge about the management choices for achieving those goals. The current implementation of VQ-ATTENDING attempts to separate these two types of knowledge as cleanly as possible and explores the design implications of using treatment goals to guide the critiquing process.

VQ-ATTENDING utilizes treatment goals as follows:

1. First, it explicitly identifies the treatment goals which it considers to apply to the patient described, for both the oxygenation and ventilation components of his respiratory management.
2. It then discusses these "active" goals in its prose critique.
3. Finally, it uses the active treatment goals internally to drive its analysis of the physician's plan.

In this way, VQ-ATTENDING deals explicitly with treatment goals both in its internal processing and in its prose critique.

The goal-directed approach to critiquing highlights several interesting system design questions: (1) How might a system best handle conflicting goals? (2) What might a system do if the physician's treatment goals differed from those of the system? (3) Could the system critique the physician's treatment goals as well as his management choices? Chapter 5 discusses these issues in more detail.

An intriguing question is why the explicit use of treatment goals is highlighted so strongly in the domain of ventilator management, while it is not so clearly evident in the other domains we explored. This question is discussed in both Chapters 5 and 9. One reason is the less structured nature of ventilator management choices, where many choices must be selected from a *continuum* (e.g., a ventilator setting might range from 0.2 to 1.0, or from 0 to 40). In contrast, with ATTENDING and HT-ATTENDING, choices are all discrete (e.g., to use drug A, drug B, or drug C). As a result, choices in ventilator management are fuzzier, which may in part explain the need to fall back to more strategic knowledge.

Nevertheless, treatment goals are clearly important in all domains of medical management. The ventilator management domain therefore highlights an important critiquing design concept. In addition, by directing attention to treatment goals, the domain yields useful insights into the process of medical management itself. In the ventilator management domain, the design implications of a goal-directed critiquing approach can be explored. Once these are understood, they may be applicable to other domains in interesting ways.

1.1.4. PHEO-ATTENDING (Pheochromocytoma Workup)

The fourth medical domain explored is very different from the three described above. Whereas the previous systems have dealt with areas of medical *management*, PHEO-ATTENDING operates in a domain of *workup*. In particular, it critiques the laboratory and radiologic workup of a patient for a suspected pheochromocytoma.

Workup is an area of medicine that lies *between* differential diagnosis and treatment. In performing differential diagnosis, a physician gathers information about a patient (symptoms, signs, and initial laboratory data) and deduces a differential diagnosis: a set of possible diseases that the patient may or may not have.

Workup involves ordering a sequence of more specific, and frequently more expensive, tests and procedures to rule in or rule out a diagnosis, or to evaluate its character and severity. Thus, workup is a second diagnostic stage. The underlying process of workup, however, is very different from that of differential diagnosis. Differential diagnosis is frequently very open ended and loosely structured and may invoke a wide range of diverse information and knowledge.

Workup, on the other hand, is very structured and focused. Workup starts in a restricted, well-defined context: a possible diagnosis to be explored further. To investigate this focused problem, there are a limited set of tests that may be used. Furthermore, in most workup domains, the rules governing the appropriate sequence of tests are also very constrained.

The total amount of knowledge required should therefore be quite manageable in many workup domains. As a result, workup should be a very fertile field for the development of practical critiquing systems. (Also, since the tests and procedures are often quite expensive, optimizing workup is economically important in helping to control the spiraling costs of health care.)

Indeed, PHEO-ATTENDING's knowledge base has proved to be significantly smaller even than that of HT-ATTENDING. Because of the rarity of pheochromocytomas, however, the system's knowledge could not be easily validated with test cases, nor would a practical consultation system be frequently used. As a result, we decided not to refine this particular system further. Nevertheless, pheochromocytoma workup proved a good domain for exploring the design issues of critiquing workup. A research project is currently in progress to apply this design to more commonly encountered workup problems.

An interesting discovery made in implementing PHEO-ATTENDING was that domain experts themselves disagree as to the optimal workup sequence. In retrospect, this problem of *conflicting expertise* is not surprising and, indeed, exists throughout medicine, as discussed in Chapter 6. Rather than try to ignore this problem, the PHEO-ATTENDING project confronted it directly and made conflicting expertise a central research focus.

Two separate PHEO-ATTENDING systems were implemented, each embodying a different expert approach. Then several experiments were performed, merging parts of the two systems into a single critiquing system that embodied both expert opinions. These experiments allowed us to explore systematically the design implications of conflicting expertise. Depending on the domain and on the type of system desired, it became clear that different merging strategies could be employed, resulting in combined critiquing systems of different character.

1.1.5. Critiquing Differential Diagnosis

There are three major clinical areas of medical practice: differential diagnosis, workup, and medical management. The four systems described in this book explore the critiquing approach in areas of management (ATTENDING, HT-ATTENDING, and VQ-ATTENDING) and workup (PHEO-ATTENDING).

It is worth emphasizing, however, that critiquing can also be applied to differential diagnosis. In fact, the ICON system (P. L. Miller et al. 1985) is currently in the early stages of implementation to critique radiologic differential diagnosis.

To use ICON, a radiologist describes the findings he sees in a chest x-ray, together with a proposed diagnosis. ICON then discusses *why* the information supplied serves to confirm or to rule out that diagnosis. ICON may also suggest further findings that can help clarify the diagnosis, again explaining why the findings are important.

The ICON system is mentioned here, and also in the discussion of future research directions in Chapter 10, because we believe that differential diagnosis will prove to be an important application area for critiquing. The system is not described in detail in this book, however, because it is still at an early developmental stage.

1.1.6. ESSENTIAL-ATTENDING (Tools for Building a Critiquing System)

During the implementation of the critiquing systems described above, we have also developed domain-independent tools that can be used to build critiquing systems in further domains. These tools are incorporated into ESSENTIAL-ATTENDING (E-ATTENDING), a critiquing system-building system. A major advantage of using such a system is that it provides the system builder with an existing design.

Several previous expert system-building systems have been developed for domains of medicine. These included E-MYCIN (van Melle 1979), EXPERT (Weiss and Kulikowski 1979), and KMS (Reggia and Pericone 1981). None of these, however, is designed to help implement a *critiquing* system.

In developing E-ATTENDING, a special effort has been made to keep the system as simple as possible. The current E-ATTENDING system can be used as it stands to build critiquing systems in a fairly wide spectrum of interesting medical domains. (HT-ATTENDING, VQ-ATTENDING, ICON, as well as a system being developed to critique the workup of obstructive jaundice, are all currently implemented using E-ATTENDING.)

At the same time, however, there are a number of more sophisticated critiquing capabilities that the current E-ATTENDING does not provide. An advantage of keeping E-ATTENDING simple is that it can be completely understood by an interested system designer, who can then add further capabilities if desired.

1.1.7. Schematic Overview

This section concludes the introductory overview of this book. Figure 1.2 summarizes schematically the various critiquing systems described. As the figure indicates, we are currently implementing critiquing systems in all three major areas of medicine: medical management, workup, and differential diagnosis.

1. ATTENDING, VQ-ATTENDING, and PHEO-ATTENDING are research prototype systems implemented to explore basic critiquing design issues. These are not envisioned as practical consultation systems in the near future.
2. Although HT-ATTENDING was initially developed as a research prototype, a major current priority is to refine that system for use as a practical consultation tool.

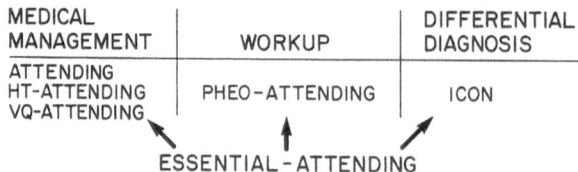

Figure 1.2. A schematic overview of the various critiquing systems described in this book.

3. In addition, although pheochromocytomas are too rare to make further refinement of PHEO-ATTENDING worthwhile, we do anticipate that other domains of workup may yield practical consultation systems in the near future.
4. Finally, the domain-independent ESSENTIAL-ATTENDING system is already being used to help implement critiquing systems in all three of these medical areas.

1.2. Critiquing: An Introductory Example

To help make the concept of an expert critiquing system more concrete, this section shows an introductory example of such a system in operation. In this example, the ATTENDING system critiques a plan for anesthetic management.

To use ATTENDING, a physician first describes a patient and a proposed surgical procedure. He does this by a process of "menu selection," as shown below. [In the computer output that follows, the input typed by the physician is underlined. Explanatory text is enclosed in square brackets.]

```
***  please describe your patient
***  has this patient suffered acute trauma or hemorrhage?
     type Y or N ***N
***  do you consider the patient to have a full stomach?
     type Y or N ***N
***  is the patient potentially hypovolemic?
     type Y or N ***N
```

[ATTENDING first asks three basic yes-no questions and then displays a menu of 23 further medical problems whose anesthetic implications are included in the system's knowledge base. These include a good number of the anesthetically most important and interesting problems. Since a patient is allowed to have any combination of these problems, ATTENDING can handle quite a wide variety of challenging patients. This set of medical problems, however, is far from complete. This restriction in its knowledge of underlying medical problems is ATTENDING's biggest current limitation.]

```
***  please specify any other problems from this list:
      1. none
      2. asthma
      3. liver disease
      4. chronic renal failure
      5. high blood pressure
      6. a prior stroke
      7. coronary artery disease
      8. a penetrating eye wound
      9. an abdominal aortic aneurysm
     10. an esophageal hemorrhage
     11. increased ICP
     12. a difficult airway
```

13. facial trauma
14. an unstable neck
15. a cerebral aneurysm
16. carotid disease
17. coagulopathy
18. myasthenia
19. hyperkalemia
20. morbid obesity
21. a bowel obstruction
22. COPD
23. a history of MAOI treatment
type one or more numbers separated by spaces *** 2 5 11

[Here the physician has described a patient with asthma, a history of high blood pressure, and increased intracranial pressure (ICP) who is coming to surgery on an emergency basis with a full stomach. Next, the physician is asked to specify a surgical procedure, again by menu selection.]

*** please indicate a surgical procedure:
1. appendectomy
2. an abdominal operation
3. an abdominal aortic aneurysm
4. a carotid endarterectomy
5. cholecystectomy
6. colectomy
7. craniotomy
8. an operation on the head or neck
9. a lower extremity operation
10. tendon repair
11. a total hip repair
12. an upper extremity operation
select one ***7
history: (ASTHMA HYPERTENSION INCREASED-ICP FULL-STOMACH)
operation: CRANIOTOMY
type 0 if ok . . . or else, type 1 to change ***0

[Once the physician confirms that he has described the patient and procedure correctly, he next outlines agents and techniques for the patient's anesthetic management. As shown below, a plan for general anesthesia involves four steps: (1) *premedication*, (2) *induction* (putting the patient to sleep), (3) *intubation* (inserting a breathing tube into the patient's trachea to secure and protect his airway), and (4) *maintenance* (keeping the patient asleep). Alternatively, a plan for regional anesthesia may be proposed (e.g., spinal or epidural anesthesia).]

— PLEASE OUTLINE A PLAN —

*** PREMED: 1-NONE   2-ATROPINE   3-GLYCOPYRROLATE
4-MORPHINE   5-DEMEROL   6-FENTANYL   7-VALIUM

```
    8-PENTOBARBITAL   9-SECONAL   10-VISTARIL
    11-DROPERIDOL   12-PHENERGAN
    type one or more numbers separated by spaces ***2 4

*** INDUCTION: 1-THIOPENTAL   2-KETAMINE   3-HALOTHANE
    4-ENFLURANE
    select one ***1

*** INTUBATION: 1-NONE   2-CONVENTIONAL   3-RAPIDSEQ
    4-AWAKE-ORAL   5-AWAKE-NASAL   6-MASKCRICOID
    7-SUCCINYLCHOLINE   8-CURARE   9-PANCURONIUM
    10-FLAXEDIL   11-METUBINE
    type one or more numbers separated by spaces ***3 7

*** MAINTENANCE: 1-HALOTHANE   2-ENFLURANE   3-N2O   4-O2
    5-MORPHINE   6-DEMEROL   7-FENTANYL   8-CURARE
    9-PANCURONIUM   10-FLAXEDIL   11-METUBINE
    type one or more numbers separated by spaces ***1 3 4

*** you have selected:
    1. general
    2. premedication—ATROPINE MORPHINE
    3. induction—THIOPENTAL
    4. intubation—RAPIDSEQ SUCCINYLCHOLINE
    5. maintenance—HALOTHANE N2O O2
    type O is ok . . . else number for change ***0
```

[Once the physician confirms that his anesthetic plan has been input correctly, ATTENDING then produces its prose critique of this plan for this patient's management.]

"There are a number of basic principles involved in this patient's management. First, in a patient with a full stomach, one must consider the problem of aspiration. Second, a patient with asthma must be protected from bronchospasm. Third, in the presence of high blood pressure, one must attempt to prevent extremes of hypertension. Fourth, in a patient with increased ICP, one must consider the problem of further increase of ICP.

"Intubation of this patient is certainly necessary because of a full stomach. Rapid-sequence intubation is a reasonable approach. This, however, involves the conceivable risk of bronchospasm. The choice of succinylcholine might have the risk of slight transient increased ICP. An alternative would be high-dose pancuronium. An alternative approach to rapid sequence intubation would be intubation after mask induction with cricoid pressure.

"From the standpoint of maintenance, inhalational technique has the advantage of giving better control of blood pressure, and of preventing bronchospasm by keeping the anesthetic level deep. On the other hand, it could have the risk of increased cerebral blood flow which could, however, be counteracted by hyperventilation. An alternative approach, perhaps, to inhalational technique would be balanced anesthesia. This would have two possible risks. First, there is the risk that this could cause possible bronchospasm due to light anesthesia. Second, in a patient with high blood pressure, this may not give adequate control of hypertension. It does, however, have the advantage of reducing cerebral blood flow.

"Looking at other aspects of the proposed plan, for a patient with increased ICP, induction with thiopental has the advantage of decreasing cerebral blood flow.

"In addition, there are several further considerations worth mentioning.
"1. Narcotic premedication involves the risk of increased ICP due to hyper-carbia.
"2. Using no relaxant for maintenance might have the risk of inadequate sup-pression of movement.
"3. Morphine premedication could have the conceivable risk of broncho-spasm due to histamine release.
"4. Induction using thiopental would have the conceivable risk of bron-chospasm."

[This concludes ATTENDING's critique.]

The prose critique shown above is designed to give the physician feedback to help him refine his plan. It has been said of anesthetic management, "It is not as important what you do, as that you do it well." This implies that an anesthetist must think through his plan to minimize possible risks and prepare ahead of time for any problems that may be anticipated to occur. (Some approaches, of course, may be clearly superior to others if major risks can be avoided.)

ATTENDING attempts to make sure the physician has thought through the implications (the risks and benefits) of different approaches to a patient's anesthetic management. The physician is then free to do as he pleases.

The following comments may help put the critique shown above into per-spective.

1. The paragraphs of the critique are not "canned" text. They are pieced together from individual words and sentence fragments, tailored to the partic-ular plan to the particular patient's care.
2. ATTENDING's introductory paragraph is a fairly primitive attempt to put the body of the critique into context by discussing the underlying management principles involved. In the "goal-directed" VQ-ATTENDING system, this problem is dealt with in a more sophisticated way. In that system, a set of individualized treatment goals are inferred. These goals are then discussed in the critique in a introductory fashion and are also used internally to drive the critiquing analysis.
3. ATTENDING's critique focuses on alternative management choices, with a fairly superficial discussion of the risks involved in each. ATTENDING men-tions each risk but does not discuss it in detail. For some risks, a superficial discussion may be all that is needed. For others, however, the system may use-fully include more detailed discussion of underlying mechanisms, of why the

risk is important, of how it is best avoided, etc. In the other systems described in this book, the system design was modified to allow more descriptive, explanatory material of this sort to be included in the critique.

1.3. Rationale for the Critiquing Approach

Now that we have shown an introductory example of the critiquing process, this section discusses *why* critiquing may be useful. This question can be approached from several perspectives.

1. The clinical perspective. The primary motivation for exploring critiquing stems from a conviction that it brings advice to the physician in the form he can best use. As discussed below, the large amount of *variation* and *subjectivity* inherent in medical practice makes it virtually impossible to try to tell a physician what to do.
2. Critiquing as a form of practice-based education. A major advantage of the computer is that it can remember large amounts of detailed information, whereas the human tends to forget information rapidly, especially if the information is seldom used. The critiquing approach allows the computer to structure detailed information around a physician's daily practice. As a result, critiquing is not only a form of consultation but simultaneously serves as a form of practice-based continuing medical education.
3. The computer science perspective. A final perspective concerns the contribution of critiquing to the field of expert systems research. From this standpoint, critiquing can be seen as a way to structure *explanation* by an expert system of its internal logical processes, as discussed below.

The remainder of this section discusses critiquing from each of these perspectives. Before we proceed, however, an anecdote may help motivate our rationale for advocating the approach.

A computer scientist (not the author) was recently working with a group of anesthetists to develop decision-making aids. One problem he wanted to model was the choice of appropriate premedication prior to surgery. He was having great difficulty getting the anesthetists to articulate their reasons for deciding (1) whether or not to premedicate a patient, and (2) what premedication to use.

When he tried to press the anesthetists for their reasoning, different criteria were given (how anxious the patient was, the type of operation, etc.) but frequently it seemed that more intuitive judgements were being made. He was concerned that a host of subconscious, internalized decision rules existed, which might prove very difficult to elicit and to model.

In retrospect, it is clear that there is *not* a host of fixed, subtle rules for choosing appropriate premedication, but rather that for most patients, there is totally *free variation*. For most patients, it does not matter whether premedication is given or not, nor does it matter what premedications are used. In short, the choice of premedication is akin to the choice of coffee, tea, or milk with breakfast. One can do as one pleases.

On the other hand, for a small number of patients, decisions concerning premedication are indeed important. For example: (1) a patient with asthma, with a cerebral aneurysm, or with an aortic aneurysm probably *should be* premedicated. (2) A patient with severe respiratory disease or a large airway mass probably *should not be* premedicated. Also, a variety of diseases are relative contraindications to specific premedication agents. For example: (1) a patient with increased intracranial pressure should not be given a narcotic. (2) A patient with Parkinson's disease should not receive a phenothiazine derivative.

The patients for whom premedication decisions are medically important, however, represent only a small fraction (well under 10%) of patients coming to elective surgery. For all the rest, a physician can safely do as he prefers.

As a result, if a system builder tries to design a system that recommends appropriate premedication for all patients, he is pursuing a mirage. There is too much latitude for individual variation and subjectivity. It makes much more sense to let the physician indicate what he would like to do, and to give him feedback only if this involves particular risks or benefits. This is what the critiquing approach is all about.

1.3.1. Clinical Advantages of Critiquing

Our primary motivation for advocating the critiquing approach stems from a conviction that it brings computer-based advice to the physician in the form he can best use. There are a number of reasons for this conviction.

1.3.1.1. Practice Variation. There is great latitude for practice variation throughout medicine. There are frequently several ways to approach a problem, and it is seldom the case that one approach is "right" and the others are "wrong." In fact, frequently there is "conflicting expertise" where even domain experts disagree as to the best way to approach particular problems.

Practice variation has a number of sources:

1. Different institutions may advocate idiosyncratic styles of practice. As a result, a doctor's training, which typically takes him through several institutions, leaves him with a unique blend of the approaches to which he is exposed.
2. Medical practice is constantly in flux. New techniques and new knowledge diffuse at different rates into practice. Some of the new approaches flourish, others pass into obscurity.
3. A physician therefore adapts a style of practice that reflects a host of variables to which he is exposed, including his personal experience of what has worked well for him in the past and what has not.

In the face of all this variation, it is hard to imagine a computer telling the physician what to do.

1.3.1.2. Subjective Evaluation. In addition to practice variation, medical decision making is further complicated by the highly subjective nature of medical practice. In making treatment decisions, a great deal of subjective judgment is often used in assessing the severity of a patient's disease.

The simple decision of when a patient with chronic illness merits hospitalization is a good example. Many factors are taken into consideration: the severity of the patient's symptoms, the patient's prior experience with similar symptoms, the ability of the patient to look after himself, the support available at home, etc. Such a decision cannot easily be dictated even by a complex set of decision rules. It requires human judgment and common sense.

Such decisions abound in medicine. In treating hypertension in a patient with congestive heart failure (CHF), a physician may consider using a beta blocker only if the CHF is "well compensated," a subjective assessment. In assessing risk tradeoffs in anesthetic management, an anesthetist may weigh the severity of, say, a patient's coronary artery disease against the severity of his asthma. Even though one might attempt to quantify such a decision, the ultimate choice must be based upon an informed subjective evaluation.

Indeed, different physicians may subjectively evaluate a particular patient differently and may therefore be led to a different approach. Even if the computer could fully model the judgment of one physician (an impossible task for the current state of the art), this model would be of limited general value.

It is therefore better to design a computer to outline the various reasonable options and help the physician think the problem through for himself. By helping optimize his plan, the computer leaves the physician in primary control of patient care, where indeed he belongs.

1.3.2. Critiquing as Practice-Based Continuing Medical Education

A second motivation for adopting the critiquing approach is that it allows education to be mixed with consultation in a natural way. A major advantage the computer has over the physician is the accuracy and capacity of its memory. The computer can remember a vast amount of complex, detailed information. Humans are notoriously poor at this task.

For example, in building the ICON system that critiques radiologic differential diagnosis, our domain experts combed the radiologic literature for detailed diagnostic criteria for a fairly rare condition. Months later, ICON would include this information in its analysis of a case. By this time, the domain experts themselves had forgotten many of the details. One radiologist who had actually assembled most of the information commented jokingly, "Whoever wrote this (critique) sounds like he really knows what he's talking about!"

Thus, a critiquing system can be more than a consultation tool. Since it can bring complex clinical information to the physician structured around a particular case, it provides a potent vehicle for practice-based continuing medical education.

1.3.3. Critiquing: The Computer Science Perspective

The previous sections have discussed possible clinical advantages of the critiquing approach. This section examines critiquing from the computer science standpoint. What contribution does the research described in this book make to the field of expert systems research?

From the computer science perspective, critiquing can be seen as a form of *explanation* by an expert system of its own internal logical processes. Explanation is an active field in expert system research. This research area has evolved from a conviction that any recommendation the computer makes will be of little value to a physician unless he is satisfied that it makes sense.

A recent study (Teach and Shortliffe 1981) highlighted this problem. A group of physicians was queried regarding their attitudes to computer consultation. An interesting finding was that the physicians did not insist that the computer always be "right." Apparently, they felt this was too much to ask of any expert, human or machine. Also, as discussed above, it is often difficult to determine absolute "rightness" in any case.

The physicians did feel it was most important, however, that the system be able to *explain and justify* its recommendations. Only if the computer could explain why it had come to its conclusions would the physician be able to determine whether they indeed made sense for his patient.

A number of research projects have explored the implementation of *explanation capabilities* for expert systems (Swartout 1981; Scott et al. 1984). One approach has been to use "rule-based" explanation. Here, the program's domain knowledge is embedded in "production rules," IF-THEN constructs used by the system to make inferences. In coming to a conclusion, a chain of such inferences is built up. The system can then explain its conclusions by tracing back through the chain of inferences which has led to that conclusion.

Such a system can provide a step-by-step explanation that exactly mirrors the system's internal logic. Further research is in progress to augment the scope of such explanation in various ways. One augmentation would be to incorporate more strategic knowledge, e.g., about diagnostic goals. (The goal-directed design of VQ-ATTENDING deals with this problem.)

A limitation of a rule-driven explanation capability is that, while it is good at answering "why" questions, it is less well designed to answer "why not" questions. ["Why not" questions have been dealt with in such systems, but in a limited way (Davis and Lenat 1982).] As a result, these systems are good at explaining their choices but less oriented to explaining why other choices were not made. The critiquing approach, of course, directly addresses "why not" questions.

Because of the great practice variation inherent in medicine, "why not" questions are particularly important. An example, in fact, occurred with a system that recommends anesthetic management (Harrison and Johnson 1981). When this system was subjected to clinical trial, one of the major questions users had was *why not* do things in their accustomed fashion.

The traditional explanation capability allows the physician to adapt an expert system's advice to the physician's thinking, *using the computer's advice as a start-*

ing point. Critiquing turns this paradigm around. Instead of starting with the computer's recommendation, a critiquing system first elicits the physician's plan, which then becomes the starting point of the analysis. In this way, a critiquing system focuses its advice around the physician's thinking in a direct and natural way.

1.4. Rationale for a Research Strategy: Prototype Systems

Our initial research strategy in exploring the critiquing approach has been to develop *several prototype systems* in different medical domains. A reader unfamiliar with computer science research may find this research strategy unusual. Much basic computer science research, however, involves the implementation of prototype systems to explore complex design issues.

An alternative approach might have been to focus on a single domain, such an anesthetic management, the first domain we explored. We might have devoted our efforts to extending and refining ATTENDING. We have already discussed some reasons for not pursuing that domain in this fashion. The advantages of our strategy include:

1. Certain facets of critiquing are seen more clearly in different domains.
2. A number of common features can be identified that apply in many domains.
3. Exploring several domains yields a better sense of domain characteristics that make the critiquing approach successful.
4. It also gives us a feel for domain characteristics that may facilitate the implementation of *practical* critiquing systems.

The remainder of this section discusses each of these issues in turn.

1.4.1. Certain Facets Are Seen More Clearly in Particular Domains

The primary goal of this work is to explore the design issues involved in implementing the critiquing approach. As outlined in Section 1.1, a number of interesting issues were much more apparent in particular domains, for instance:

1. The implementation of ATTENDING (anesthetic management) highlighted the need to deal explicitly with *risk* and to develop a heuristic approach to risk analysis.
2. Since the HT-ATTENDING system was modeled after a survey paper on hypertension management, the design of that system showed how a critiquing system could closely mirror such a paper and, indeed, might be perceived as a highly interactive "paper" that tailored its content to a particular patient's care.
3. The implementation of VQ-ATTENDING (ventilator management) showed how an explicit assessment of *treatment goals* could be made a central component of a critiquing system's analysis.

4. The PHEO-ATTENDING implementation (pheochromocytoma workup) highlighted the importance of *conflicting expertise* and how an expert system might incorporate conflicting expertise in different ways.

Thus, the initial exploration of several domains has yielded a more comprehensive understanding of the critiquing process. An attempt to formulate some of the general lessons learned from this experience is outlined in Chapter 9.

1.4.2. Commonality

Another advantage of exploring several domains is that certain features have emerged as common to many, if not all. We have attempted to capture this commonality in the ESSENTIAL-ATTENDING system-building system, as described in Chapter 7. This system provides a formalized skeleton to assist in building critiquing systems in further domains.

1.4.3. Identifying Appropriate Domains for Critiquing

Some domains will presumably prove better suited for critiquing than others. If there is only one way to approach a particular problem, then critiquing will have little to offer, unless the approach is so complex that it is difficult to remember. Similarly, if there are several choices but no particular risks or benefits (i.e., if there is totally free variation), here again little would be gained from a critiquing system.

A critiquing system is probably best suited to domains where:

1. There are a number of alternative choices.
2. There are a number of different risks and benefits associated with the various choices in different patients.
3. New treatment choices and new knowledge about existing features periodically alter the field.

Having implemented critiquing systems in several domains, it is our feeling that many, if not most, domains of medicine have these characteristics, and that the critiquing approach will therefore prove widely applicable.

1.4.4. What Makes a Domain Good for Practical Consultation?

Another major advantage of exploring several domains is that it helps us better understand the characteristics that make a domain ripe for the development of a *practical* critiquing system.

1.4.4.1. Sufficiently Constrained Domains. Perhaps most important is that a domain be sufficiently constrained. In the whole field of AIM research, there are only three systems currently in routine clinical use. All three are in domains that are very constrained. Two of these, SPE (Weiss et al. 1978), which interprets serum protein electrophoresis data, and PUFF (Aikins et al. 1983), which interprets pulmonary function tests, are constrained in that they each interpret the results of a laboratory instrument. (Indeed, since these systems do not interact with the physician, they are not true consultation systems.) The third system in routine use is ONCOCIN (Shortliffe et al. 1981) which assists in the implementation of oncology protocols. This system is constrained in that it implements existing protocols that specify in detail how chemotherapeutic drugs are to be administered.

Figure 1.3 illustrates this problem schematically using a diagram initially formulated by Blois (1980) in an article about computer-assisted medical decision making. The funnel-like lines illustrate the *scope* of different medical domains for which computer advisors may be built. Domains that fall in the open area at the left are very broad and unconstrained. A great deal of domain knowledge would be required to implement a system successfully in these domains. Domains that fall in the narrow area at the right are tightly focused and highly constrained.

Figure 1.3 shows roughly where we feel the four systems described in this book fall along this spectrum. Essential hypertension (HT-ATTENDING) is the most complex domain in which we currently feel comfortable pursuing a practical system. Domains that are less constrained, however, may still be productive areas for basic research into more sophisticated critiquing capabilities.

1.4.4.2. Frequency of the Medical Problem. A further characteristic that makes a domain good for practical consultation is the frequency of the clinical problem. Thus, HT-ATTENDING's domain, essential hypertension, is promising because of the prevalance of the disease. On the other hand, the medical problem dealt

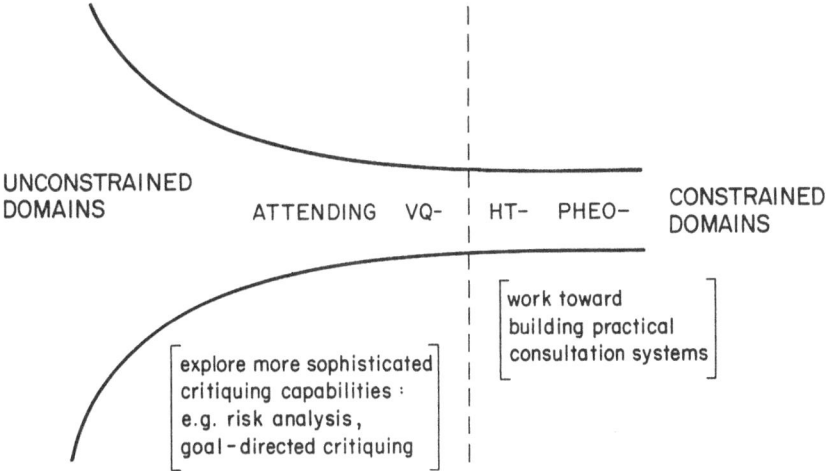

Figure 1.3. A schematic diagram outlining how certain domains may be very broad, while others may be very constrained.

with by PHEO-ATTENDING (a suspected pheochromocytoma) is so rare that it would be difficult even to get sufficient cases to test the system.

1.5. Where We Go from Here

Where, then, might critiquing research go from here? Certainly, it eventually becomes unproductive to go from one domain to another, implementing only prototype systems. At some point, we must focus on one or more domains and explore the problems of making a practical consultation system work. At the same time, however, we also plan to continue basic research on more sophisticated critiquing capabilities.

Thus, there are two general future directions for critiquing research: (1) developing practical consultation systems in constrained domains, and (2) pursuing basic research on sophisticated critiquing capabilities. Some of this basic research may be done in domains where practical systems are not feasible.

As mentioned previously, we feel the HT-ATTENDING system is ripe for development as practical tool. In addition, we anticipate that a number of constrained domains of workup will prove good areas for the development of practical systems. At the same time, there are a variety of ways in which the critiquing approach can be made more sophisticated. The work described in this book is a step toward exploring these directions.

Chapter 2

Artificial Intelligence in Medicine (AIM)

The development of expert critiquing systems is part of a growing field involving numerous projects applying artificial intelligence in medicine. This work has been in progress for the past 15 years (Clancey and Shortliffe 1984; Kulikowski 1980; Shortliffe et al. 1979; Szolovits 1982). This chapter gives an overview of these projects. It then discusses the field of AI as a whole. Finally, the chapter discusses how critiquing itself can be performed at different levels of complexity. The critiquing research described in this book focusses on the more complex end of this spectrum.

2.1. Previous AIM Research

There are a number of perspectives from which one might discuss previous AIM projects.

1. Diagnosis vs. management. Most of the initial AIM work dealt primarily with medical diagnosis. More recently, however, increasing numbers of systems have explored various domains of medical management (treatment).
2. Cognitive models of physician behavior. Another perspective is whether a system attempts to model a physician's cognitive behavior. Most systems have focused on the difficult problems of trying to make the computer perform appropriately, without necessarily attempting to model the physician's cognitive processes.
3. Size of the domain. Certain systems have tackled very large, ambitious domains. More recently, there has been a trend toward smaller, more constrained domains.
4. Practical system use. Most previous AIM systems have been developmental prototypes exploring basic design issues but have not led to practical systems. More recently, a small number of working systems have emerged.

5. Internal knowledge manipulation. All AIM systems have the primary goal of assisting in clinical decision making. In addition, however, several projects have explored how a system can "introspectively" manipulate its domain knowledge for other purposes. These other purposes include: (1) explanation; (2) teaching; (3) verification of the accuracy, completeness, and consistency of its knowledge; and (4) assisting in the acquisition of new knowledge from the domain expert.

The remainder of this section discusses several research projects from these perspectives. This overview is organized around four major centers of AIM research during the past 15 years: Stanford, Rutgers, MIT, and the University of Pittsburgh.

2.1.1. AIM Research at Stanford

The earliest expert consultation system developed at Stanford in the late 1960s, DENDRAL, does not deal with clinical medicine but helps interpret nuclear magnetic resonance data in identifying organic compounds (Lindsay et al. 1980). This system served as the father of much of the later work. In particular, DEN-DRAL explored the use of the production rule formalism for representing domain knowledge.

Production rules were later used in MYCIN (Shortliffe 1976), a system that dealt with both diagnosis and treatment of infectious disease. MYCIN was the starting point for a cluster of related research projects, most of which explored different aspects of the production rule approach (Buchanan and Shortliffe 1984). TEIRESIAS (Davis and Lenat 1982) explored the use of MYCIN's knowledge base both for explanation and to assist in the acquisition of new knowledge. GUIDON and NEOMYCIN (Clancey and Letsinger 1981; London and Clancey 1982) explored the use of MYCIN's knowledge for teaching. VM (Fagan et al. 1979) extended the rule-based approach beyond diagnosis into a domain of management. The domain chosen, ventilator management, was particularly interesting since it involved monitoring a patient's clinical status over time and therefore required developing an approach to time-oriented reasoning.

A later system, ONCOCIN (Shortliffe et al. 1981), also applied the production rule approach to management, helping implement oncology protocols for the administration of chemotherapy. ONCOCIN was one of the first AIM systems to be brought into routine use, in Stanford's oncology clinic. PUFF (Aikins et al. 1983), developed in collaboration with Stanford, interprets pulmonary function tests and is in routine use at the Pacific Medical Center.

2.1.2. AIM Research at Rutgers

An early system developed at Rutgers, CASNET, performed ophthalmologic diagnosis (Weiss et al. 1978). From the experience gained in implementing this system, a domain-independent system-building system, EXPERT, was built (Weiss and Kulikowski 1979). Like MYCIN, EXPERT employs production rules to structure domain knowledge. Systems have been built using EXPERT in a number of medical domains. Two of these, AI/COAG (which diagnoses coagulation disorders) and AI/RHEUM (which performs rheumatologic diagnosis) were built in collaboration with researchers at the University of Missouri (Kingsland et al. 1982; Lindberg et al. 1980).

Two interesting extensions of this research are:

1. SEEK assists AI/RHEUM's designer in validating that system's knowledge (Politakis and Weiss 1984). SEEK takes AI/RHEUM's knowledge, plus a set of test cases whose diagnoses are known, and proposes modifications that might let the system perform more accurately.
2. A small diagnostic system, SPE, interprets serum electrophoresis data produced by a laboratory instrument (Weiss et al. 1981). SPE was developed using EXPERT on a large computer and was then "compiled" onto a microchip that has been incorporated into a commercial SPE analyzer. As a result, it is one of the few working AIM systems.

2.1.3. AIM Research at MIT

One of the early AIM systems at MIT was PIP (Present Illness Program) developed in collaboration with Tufts Medical School (Pauker et al. 1976). This system is interesting in that it does not merely attempt to correlate a set of findings with a diagnosis but also tries to develop a model of *how* a physician may be using his clinical knowledge when eliciting present illness information about a patient.

The Digitalis Advisor (Gorry et al. 1978) modeled an area of medical treatment and was later adapted to explain its recommendations using an "automatic programming" model for the explanation process (Swartout 1981).

ARRY (Long and Russ 1983; Long et al. 1983) monitors ventricular arrhythmias in an intensive care setting and makes both diagnostic and therapeutic suggestions. Two interesting features of this system are : (1) it incorporates a flexible representation for time-dependent data, and (2) it integrates information from several knowledge sources (including EKG, laboratory data, and pharmacokinetic models) in making its recommendations.

ABEL (Patil 1981) models the "deep," causal structure underlying acid-base analysis and demonstrates how this knowledge can be approached at different "levels of abstraction."

In addition, a system currently under development deals with the diagnosis and management of congestive heart failure (Long et al. 1982). This system is developing a "qualitative model" of the cardiovascular system and uses a "truth maintenance system" to let the system itself evaluate the effects of different therapeutic interventions.

2.1.4. AIM Research at the University of Pittsburgh

The most ambitious AIM system is INTERNIST (also called CADUCEUS), under development at the University of Pittsburgh, which attempts to embrace the whole of internal medicine. In an evaluation of the system's performance (R. A. Miller et al. 1982; Pople 1982), it was compared against clinicians on a series of clinicopathologic cases taken from the *New England Journal of Medicine*. Although the system demonstrated impressive performance (roughly comparable to that of the clinicians reported), a detailed analysis showed a number of fundamental areas where the system's reasoning could be made more sophisticated. These refinements include a better model of causal, anatomic, and temporal reasoning in the diagnostic process.

SCAN (Banks and Weimer 1983) is currently being developed to tackle one of these problems (anatomic reasoning) in the domain of neurology. SCAN has a detailed model of the central nervous system which it uses to correlate neurologic symptoms into an anatomically based diagnosis.

2.2. What Is Artificial Intelligence?

Now that several previous projects applying artificial intelligence (AI) in medicine have been described, this section addresses a fundamental question: "what is AI, and what if anything makes an AI system different from other medical computer systems?"

A precise definition of artificial intelligence is not easily articulated. AI is usually described as the science of designing a computer system so that it will respond or behave in a manner that appears to exhibit "intelligence." Several qualifications, however, must be added to this definition.

1. Any computer program that performs a purely numeric analysis, such as a payroll program, a statistical package, or a complicated scientific algorithm is not considered AI.
2. On the other hand, a number of sensory and motor skills that humans take for granted (and that animals share) involve the sophisticated utilization of real-world knowledge. Identifying objects in a visual scene, for example, requires knowledge about the objects being observed. As a result, the implementation of these skills is usually considered to be AI.

This definition outlines the overall goals of the field of artificial intelligence. It does not, however, answer the practical question of how to assess the degree of AI embodied in a given computer system. Several recent developments make this question particularly topical.

1. AI has recently become fashionable, and there is now a high level of industrial, governmental, and military interest in artificial intelligence. As a result, it is suddenly popular to identify systems as being "AI." This tendency helps spread confusion as to exactly what AI is. In previous years, AI was perceived as an esoteric, experimental field. Now that AI has become quite application oriented, the problem of differentiating it from other computer programming approaches is more apparent.

2. In addition, a small number of systems developed using artificial intelligence techniques have recently been translated into more conventional programming languages for practical use. For instance, the PUFF system, developed using EMYCIN in a LISP programming environment, was translated into BASIC to run on a microprocessor. This raises the question in some people's minds as to whether such a system is still "AI."

To understand in practice the degree to which a system involves artificial intelligence, it is helpful to ask three questions. (1) Does the system involve *basic AI research* into fundamental issues? (2) Does the system *implement sophisticated AI knowledge manipulation capabilities*? (3) Is the system *built using AI techniques*?

Figure 2.1 illustrates these three questions schematically. These criteria allow one to better understand whether, how, and why a particular system involves artificial intelligence. The remainder of this section discusses each of these three criteria in more detail.

2.2.1. Basic AI Research

Even though an exact definition of artificial intelligence is difficult to formulate, a number of basic research problems have evolved as fundamental areas of AI research. These include:

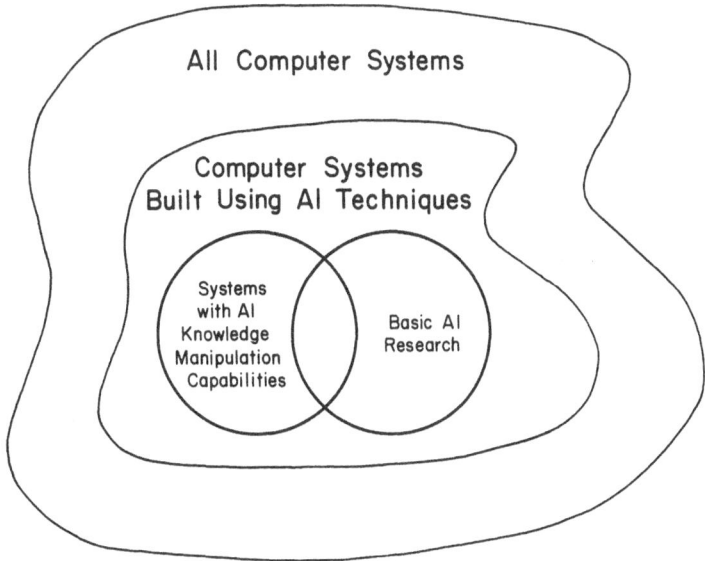

Figure 2.1. This figure illustrates that AI systems are a subset of all possible computer systems and can be characterized by three criteria.

1. Knowledge representation. To model complex real-world relationships in a flexible way, one needs sophisticated formalisms to structure domain knowledge. A wide variety of increasingly powerful techniques have been developed to deal with different types of knowledge.
2. Heuristic search. The solution to many problems can be reduced to the underlying process of searching a large "space" of possibilities. A classic example is a chess-playing program, which must explore many possible future moves. Not infrequently, the search space is so vast that heuristic methods must be developed to "prune" the search so that it can be performed in a feasible time frame.
3. Planning. A generic problem is that of planning, a central part of any complex decision-making processes.
4. Temporal reasoning. The problem of representing time, and the relationships between events which take place in time, is a difficult problem.
5. Spatial reasoning. Representing the spatial relationships between objects so a system can reason about those objects in a general way is also a nontrivial problem.
6. Reasoning from multiple sources of knowledge. A further interesting problem is developing systems that have access to a variety of different, related sources of knowledge and are able to integrate this information in an intelligent fashion.
7. Learning. A major research area involves exploring sophisticated ways in which a machine can learn new knowledge.

To the extent that a system advances the understanding of fundamental AI issues such as these, then, that system embodies basic AI research.

2.2.2. Systems That Implement AI Knowledge Manipulation Capabilities

A number of projects have explored how a system might "introspectively" examine its own knowledge in different ways.

1. Explanation. As discussed in Chapter 1, several systems have been designed with the capability of explaining and justifying their recommendations.
2. Knowledge acquisition. Other projects have explored how a computer system itself may help the domain expert enter his knowledge. This capability may be critical if the construction of expert systems is to become widespread in medicine.
3. Knowledge verification. It is similarly important that a system be able actively to assist in checking its knowledge base for accuracy, consistency, and completeness.

Most work involving knowledge manipulation capabilities such as these is still developmental and therefore currently falls into the category of basic AI research. As this technology matures, however, the incorporation of such capabilities into expert systems will become increasingly routine and will not necessarily be a hallmark of basic AI research. These systems will nevertheless be augmented with powerful capabilities that have evolved directly from AI research.

2.2.3. Systems Built Using AI Techniques

Another byproduct of basic AI research has been the development of powerful programming techniques that allow manipulation of real-world knowledge to solve complex problems. Examples of such techniques include: production rules, semantic networks, frames, truth maintenance systems, augmented transition networks, etc. An increasing number of practical systems that do not explore basic AI research issues are nevertheless built using such AI techniques. In addition to offering powerful programming constructs, the use of these techniques facilitates the incorporation of knowledge manipulation capabilities such as those described above.

2.3. A Perspective on Critiquing

A final question concerns where critiquing falls in the framework outlined above. Does critiquing require sophisticated AI techniques? Is it basic AI research? Indeed, what is critiquing? Is any computer program that gives a physician feedback a critiquing system? Just as AI spans a range of sophistication, critiquing itself involves a similar spectrum of complexity (as illustrated in Figure 2.2), ranging from systems that give simple feedback to much more sophisticated analysis. Several levels of critiquing complexity can be identified.

1. Computationally straightforward critiquing. Some systems that give a physician feedback may be computationally simple. For instance, a system may check a prescribed drug against a list of allergies by table lookup.
2. Critiquing based on objective criteria. In most domains of medicine, there is no *objective* gold standard to follow. Instead, there is usually great latitude for

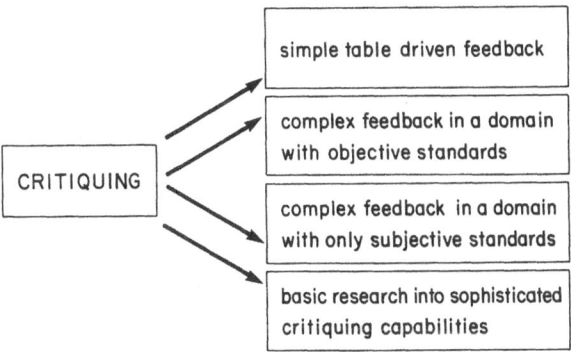

Figure 2.2. Critiquing spans a spectrum of complexity.

practice variation and subjective judgment. In certain domains, however, objective standards do exist. For example, an oncology protocol provides a detailed objective standard for patient care. In such a domain, critiquing can consist of alerting the physician when he deviates from the protocol. The ONCOCIN system, which helps implement oncology protocols, has recently been adapted experimentally to critique in this fashion (Langlotz and Short-liffe 1983).

3. Critiquing in a domain with subjective criteria. In most medical domains, however, no objective standards exist. In such domains, as described in this book, a critiquing system must lay out alternatives that it considers reasonable and discuss their relative merits. All four systems discussed in this book operate in this type of domain.

4. Basic research on sophisticated critiquing capabilities. At the more complex end of the critiquing spectrum is basic research to explore sophisticated critiquing capabilities, such as those outlined in Section 1.1.

The research described in the book focuses on (3) and (4), the more complex end of the spectrum. In fact, a more accurate name for our research might be "medical plan analysis by computer," a term that tends to emphasize a more sophisticated analysis. We use the term "critiquing" throughout this book because it is more succinct. As used in this book, critiquing implies: (1) the comprehensive analysis of a physician's plan in a nontrivial area of patient care, and (2) the creation of a comprehensive prose discussion that analyzes the physician's plan in as articulate a fashion as possible.

Chapter 3

The ATTENDING System: Anesthesiology

The implementation of the ATTENDING system, which critiques anesthetic management, was the start of our exploration of the critiquing approach (P. L. Miller 1984). ATTENDING is designed to critique an anesthetist's plan for general or regional anesthesia. A plan for general anesthesia, for instance, includes the choice of drugs and techniques proposed for:

1. Premedication
2. Induction of anesthesia (putting the patient to sleep)
3. Intubation (passing an endotracheal tube in to the patient's trachea to secure and protect his airway)
4. Maintenance of anesthesia (keeping the patient asleep)

These decisions are quite robust, in the sense that there are a number of choices for each step, choices that may be critical for sick patients. Although giving an anesthetic to a healthy patient for an elective case is usually very safe, the anesthetic management of a sick patient may involve major risks, often life-threatening risks.

Also, in a complicated patient with several medical problems, there are frequently *risk tradeoffs*. Here, techniques that are good for one problem are bad for another, and vice versa. Planning safe anesthetic management for such a patient is not simple. It is easy for the physician to focus on certain aspects of the problem without fully thinking through the implications of the patient as a whole.

3.1. Anesthesia and Risk

Anesthesia is often referred to as a "high-risk" specialty. This is a reflection of the fact that anesthetic risks are very immediate. The character of anesthetic risk, however, has undergone a dramatic change in the past 25 years. In the past, anesthetic agents themselves involved major risk. For instance, both ether and cyclopropane (anesthetic gases in common use as recently as 10 years ago) are both explosive. As a result, there was an ever-present risk of a spark igniting the oxygen-enriched gases that an anesthetized patient breathed.

In contrast, the anesthetic agents used today are remarkably safe. A small number of healthy patients may react adversely to certain anesthetic agents, but major side effects of this sort are very rare and, in any case, usually respond to treatment. There is, of course, still the potential for human error in administering anesthesia, but this too is uncommon. As a result, the administration of anesthesia to a healthy patient is very safe.

On the other hand, in recent years there has been a trend toward increasingly sick patients being brought to the operating room for anesthesia and surgery, including patients with multiorgan failure who are virtually at death's door. In contrast, 30 years ago many elderly patients, especially those with mild or moderate medical problems, would have been considered inappropriate anesthetic risks.

There are several reasons for this trend toward increasingly sick patients undergoing surgery. The dramatic increase in cardiac surgery that has taken place over the past decade has given anesthetists a great deal of experience managing unstable patients. Also, sophisticated technology is now available for these patients and is in routine use at virtually all hospitals.

As a result, although anesthesia is still a high-risk specialty, the risks today arise primarily from the underlying diseases that patients bring with them to the operating room. In planning such a patient's anesthetic management, a physician must tailor his plan to these underlying problems and to the risks they imply. In so doing, the physician attempts to reduce overall risk to the patient. The ATTENDING system is designed to help the physician in this task.

3.2. Why Anesthesia?

In retrospect, it is clear that anesthesia is an excellent domain for an initial exploration of the critiquing approach. For one thing, since risk is so central, the domain highlights risk analysis as a central underlying paradigm of medical management. This, in turn, has led to the development of a heuristic approach to risk analysis, discussed later in this chapter.

On the other hand, the domain of anesthesiology does have certain drawbacks. As discussed in Chapter 1, the amount of knowledge required to implement a comprehensive anesthesia advisor is large. As a result, the domain is not well suited to implementing a practical consultation system until the design issues have been fully explored in constrained domains. Also, anesthesia is an esoteric field. Most physicians are not familiar with anesthetic management. As a result, it may not be clear to a medical reader whether the critiquing approach as implemented in ATTENDING is widely applicable in medicine, or whether it is a reflection of certain idiosyncrasies of anesthesia.

3.3. Examples of ATTENDING in Operation

ATTENDING can be used in two modes: (1) a consultation mode and (2) a tutorial mode. In the consultation mode, as illustrated in Chapter 1, a physician describes an actual patient's underlying medical problems by a process of menu selection. Since ATTENDING is currently familiar with the anesthetic implications of only 25 such medical problems, however, the consultation use of the system is necessarily limited.

In its tutorial mode, however, the system itself presents a patient to the user (for instance, an anesthesia resident) who is asked to propose a plan for the system to critique. The hypothetical case, of course, is chosen to include only problems familiar to the system.

The examples that follow illustrate ATTENDING's tutorial mode. These examples are chosen to demonstrate certain features of the system's critique, and also to illustrate certain limitations.

Example 1. This example illustrates several facets of ATTENDING's critiquing analysis. The system starts the session by describing a hypothetical case.

> A thirty-four-year-old male with a history of asthma requiring several past hospitalizations comes to surgery after an auto accident for repair of a likely fractured spleen. His blood pressure is 85/50. He recently ate lunch.

This paragraph describes a patient with three underlying medical problems: (1) asthma, (2) hypovolemia, and (3) a full stomach, coming to surgery for an abdominal operation. The case description is stored internally as free text, with an associated list indicating the patient's medical problems and the proposed surgical procedure. This list is input to ATTENDING as if a user had described the patient. From that point on, ATTENDING's operation for consultation and tutorial use is identical.

Once the patient is described, the user is asked to outline an anesthetic plan by the same menu selection process shown in Chapter 1. Suppose, for instance, the following plan was proposed:

1. No premedication
2. Induction of anesthesia using thiopental (a short-acting barbiturate)
3. Intubation via a rapid-sequence technique using succinylcholine (a short acting muscle relaxant)
4. Maintenance using halothane (an inhalational anesthetic), curare (a muscle relaxant), nitrous oxide, and oxygen

The current version of ATTENDING produces the following critique of this plan for this patient's anesthetic management.

> There are a number of basic principles involved in this patient's management. First, in a patient with asthma, one must consider the problem of bronchospasm.

Second, a patient with hypovolemia must be protected from hypotension. Third, in the presence of a full stomach, one must attempt to prevent aspiration.

Intubation of this patient is certainly necessary because of a full stomach. Rapid-sequence intubation is a reasonable approach. This, however, could have the conceivable risk of bronchospasm. An alternative approach to rapid-sequence intubation would be intubation after mask induction with cricoid pressure.

From the standpoint of induction, induction using thiopental would have two possible risks. First, in a patient with asthma, there is the conceivable risk that this could cause bronchospasm. Second, there is the possible risk of hypotension in the presence of hypovolemia. An alternative approach to induction using thiopental would be induction with ketamine. This has the advantage of helping avoid hypotension since ketamine is supportive of blood pressure and of suppressing bronchospasm since ketamine is a bronchodilator.

In regard to maintenance, inhalational technique has the advantage of preventing bronchospasm by keeping the anesthetic level deep. On the other hand, it might have the risk of hypotension in the presence of hypovolemia. An alternative approach to inhalational technique would be balanced anesthesia. This involves the risk of possible bronchospasm due to light anesthesia. It does, however, have the advantage of maintaining blood pressure.

Looking at other aspects of the proposed plan, there are two considerations worth mentioning:

1. Curare might have the conceivable risk of bronchospasm triggered by histamine release.
2. Curare could have the risk of hypotension, again, since curare can induce histamine release.

Again, it must be emphasized that this critique is not constructed from stored paragraphs but is assembled from words and sentence fragments tailored to the particular plan for the particular patient's care. The actual logic that guides ATTENDING's analysis is described later in this chapter.

The introductory paragraph outlines the principles underlying the patient's anesthetic management and is quite rudimentary. It is not really part of the critique, since it reflects only the patient and not the proposed plan. The paragraph is designed to establish an overall context for the critique itself, which follows.

The next paragraphs discuss the various components of the plan and illustrate different types of analysis the system may perform. For instance, in the paragraph discussing induction of anesthesia, there are two possible risks of the proposed approach (thiopental) and an alternative (ketamine) that avoids these risks. As a result, the system is able to suggest an alternative that may well be superior. In contrast, in the paragraph discussing maintenance, there is a risk tradeoff. Here, an inhalational (deep) anesthetic technique is good for the patient's asthma but bad for his possible hypovolemia, whereas a balanced (light) anesthetic has the inverse risks and benefits. Thus, no approach has clear advantages. As a result, ATTENDING mentions both alternatives, outlines the risk tradeoff involved, and leaves the physician to make the final decision.

At the same time, there are many alternatives that ATTENDING has considered internally but has not mentioned in its critique. Some of these alternatives may have been clearly *inferior* to the approach proposed. For instance, *not* intubating the patient, which has the life-threatening risk of aspiration, was discarded early in the system's analysis. In addition, other alternatives, such as using the muscle relaxant metocurine rather than pancuronium, are perfectly acceptable but have no particular advantages, and are therefore ignored in the critique.

As a result, ATTENDING's critique focuses on the particular risks and risk tradeoffs involved in the proposed approach as compared to any relevant alternatives. The physician is then free to do as he pleases; he is at least alerted to the major issues involved.

Example 2. This second example involves the same patient described above and illustrates how the use of a regional anesthetic technique might be proposed. Here, the anesthetist has proposed the following plan:

1. No premedication
2. Epidural anesthesia using marcaine (a local anesthetic)

The critique produced for this plan is shorter since the regional approach involves fewer steps.

> There are a number of basic principles involved in this patient's management. First, in a patient with asthma, one must consider the problem of bronchospasm. Second, in the presence of hypovolemia, one must attempt to prevent hypotension. Third, a patient with a full stomach must be protected from aspiration.
>
> From the standpoint of using a regional technique, conduction anesthesia would have the risk of severe hypotension due to sympathectomy in a hypovolemic patient. In view of this problem with regional anesthesia, general anesthesia might be worth considering.

Notice that in this critique, the introductory paragraph is identical in its content to that of the previous critique. This is because the patient is the same, and therefore the underlying management considerations are unchanged. Notice, however, that the exact phrasing of the paragraph is slightly different. This variation in phrasing is made possible by PROSENET "option arcs" discussed in Chapter 7. Option arcs allow ATTENDING's prose output to vary and thereby helps prevent it from sounding monotonous.

Example 3. This final example involves a complicated patient with four medical problems: (1) a full stomach, (2) increased intracranial pressure (ICP), (3) morbid obesity, and (4) facial trauma. The anesthetist has suggested a plan involving:

1. Premedication with morphine and atropine
2. Induction of anesthesia using thiopental
3. Intubation using a rapid-sequence technique and succinylcholine
4. Maintenance using halothane, nitrous oxide, oxygen, and pancuronium

ATTENDING produces the following critique:

There are a number of basic principles involved in this patient's management. First, in a patient with a full stomach, one must consider the problem of aspiration. Second, a patient with increased ICP must be protected from further increase of ICP. Third, in the presence of facial trauma, one must attempt to prevent loss of an adequate airway.

Intubation of this patient is definitely appropriate because of a full stomach. Rapid sequence intubation is a reasonable approach. This, however, involves the risk of life-threatening difficulty with ventilation and intubation if the patient has a difficult airway. The choice of succinylcholine, however, might have the risk of slight transient increased ICP. An alternative would be high-dose pancuronium. An alternative approach, perhaps, to rapid sequence intubation would be awake intubation. This could have the risk of increased ICP. It does, however, have the advantage of preserving airway reflexes and respiration until the airway is secured and of securing the airway safely.

From the standpoint of maintenance, inhalational technique has the advantage of allowing high inspired oxygen. On the other hand, it would have the risk of increased cerebral blood flow, which could, however, be counteracted by hyperventilation. An alternative approach to inhalational technique would be balanced anesthesia. This has the advantage of reducing cerebral blood flow.

Looking at other aspects of the proposed plan, in the presence of increased ICP, induction using thiopental has the advantage of decreasing cerebral blood flow.

In addition, there are two further considerations worth mentioning:

1. Preoperative sedation could have the risk of depressing airway reflexes.
2. Narcotic premedication might have the risk of increased ICP due to hypercarbia.

This critique demonstrates that ATTENDING can handle a reasonably complicated patient and still produce a coherent analysis. The example also serves to illustrate certain limitations in the depth of the ATTENDING's advice.

For instance, in the intubation paragraph, ATTENDING discusses reasonably lucidly the important alternatives to consider and risk tradeoffs involved. These are discussed, however, in quite superficial terms. ATTENDING mentions each risk but does not discuss it in detail. There is a variety of information that *might* be included about the various risks, including: (1) how to assess its severity, (2) how to treat it prophylactically or guard against it preoperatively, (3) how to prepare to treat it intraoperatively, (4) the underlying causal mechanisms involved in the risk, etc. In addition, the system might go into more detail as to how best to assess and resolve particular risk tradeoffs.

As a result, there is a wide variety of information that may be added to make ATTENDING's advice more robust. The problems in adding this further information are: (1) knowing where to draw the line in such a large domain, and (2) knowing how best to organize and structure all the information to maximize its effectiveness.

As mentioned previously, we have not augmented ATTENDING in any of these ways because we felt it would be more productive to explore these issues in more constrained domains, as illustrated in Chapters 4, 5, and 6.

3.4. ATTENDING System Overview

As described previously, ATTENDING's operation is initiated by the physician describing a patient, a planned surgical procedure, and a proposed anesthetic plan, all by a process of menu selection.

The analysis of the proposed plan consists of three parts, as illustrated in Figure 3.1.

1. Exploring alternative approaches to the patient's management
2. Assessing the risks and benefits of these approaches, to determine a set of choices to be discussed in the critique
3. Generating the prose critique itself

Sections 3.5, 3.6, and 3.7 discuss in turn each of these three phases of ATTENDING's operation.

3.5. Exploring Alternative Approaches

In order to analyze the physician's plan in a comprehensive fashion, ATTENDING must explore the various possible alternative approaches to the patient's management. To allow this, ATTENDING uses the Augmented Transition Network (ATN) formalism (Woods 1970) to structure its knowledge of anesthetic management and to coordinate its analysis of the plan.

3.5.1. ATTENDING's ATNs: Modeling Anesthetic Management

The ATN formalism was originally developed for natural language processing: computer programs that "understand" English text or speech. The ATN formalism has proved a very flexible tool, however, and has been applied to a number of problems beyond natural language. It has previously been used to model decision making in the Programmer's Apprentice system (M. L. Miller and Goldstein 1977).

The easiest way to introduce ATTENDING's use of the ATN is to start with a simple example and then generalize to more complex examples. Figure 3.2 shows a very simple ATN that models two choices for *inhalational* anesthesia. The network itself is named "INHAL."

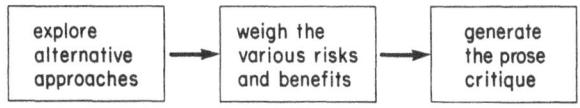

Figure 3.1. The three components of ATTENDING's operation.

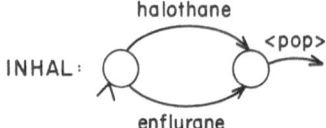

Figure 3.2. A simple ATN.

The network consists of states (circles) connected by arcs. Starting at the initial state of the network, one has a choice of two arcs, both of which lead to the same destination state. Each arc is labeled with the name of an anesthetic technique, in this case, the inhalational anesthetic agents "enflurane" and "halothane." The process of tracing a path through the network, from state to state, corresponds to the process of making anesthetic management decisions. (A *pop* arc indicates the end of the path through the network.)

As illustrated in Figure 3.3, each arc also has an associated list of risks and benefits. These correspond to the risks and benefits of using that anesthetic technique in the presence of different medical problems. (Exactly how these risks are handled is discussed more fully in Section 3.6.)

For example, the list of risks and benefits associated with the ENFLURANE arc would include:

1. That enflurane has a theoretical risk of renal toxicity in a patient with renal failure
2. That enflurane has the risk of initial bronchospasm in a patient with asthma
3. That enflurane has a risk of causing increased seizure activity in a patient with increased intracranial pressure

Using these risks associated with the different arcs, ATTENDING can explore an ATN network searching for paths (anesthetic approaches) that minimize risk.

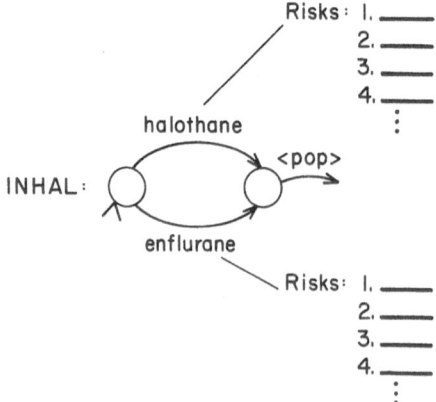

Figure 3.3. Each ATN arc has an associated list of risks and benefits.

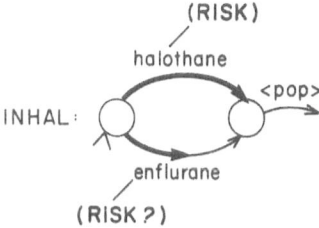

Figure 3.4. Exploring alternative paths through the simple ATN.

For example, as illustrated in Figure 3.4, if the physician has proposed the use of halothane (the dark path) and halothane involves some risk for the patient, then ATTENDING can examine other paths to see if they may involve less risk. (In this simple example, of course, there is only one other path to consider.)

Figure 3.5, on the other hand, shows a more complicated set of networks. The dark line through these networks corresponds to a completely formulated anesthetic plan:

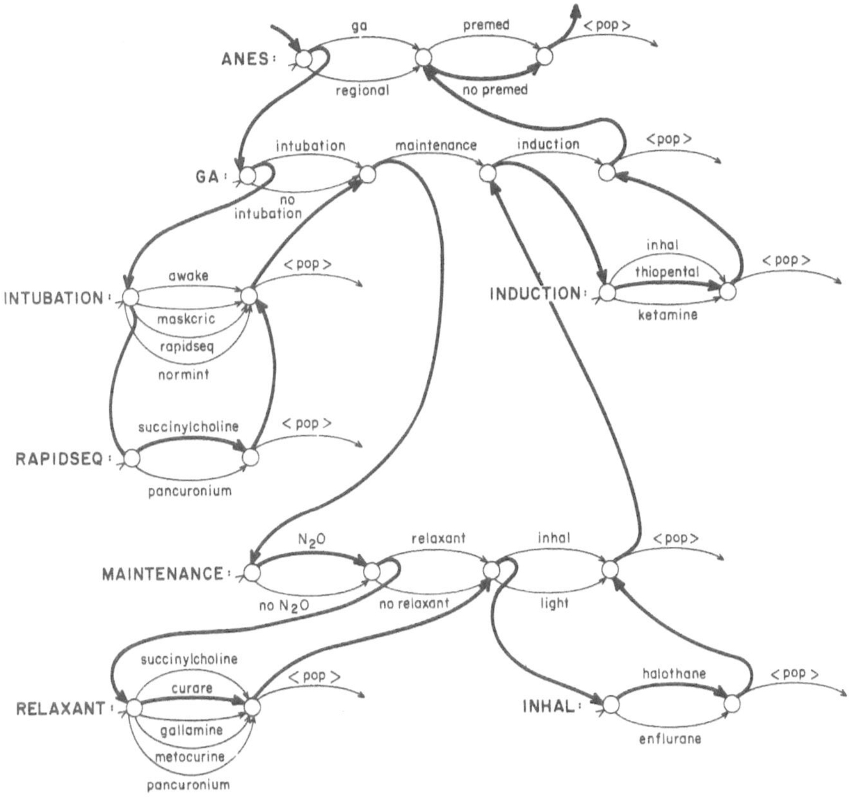

Figure 3.5. A path through the entire set of ATN networks represents a completely formulated anesthetic management plan.

1. No premedication
2. Induction using thiopental
3. Rapid-sequence intubation using succinylcholine
4. Maintenance of anesthesia using halothane, pancuronium, nitrous oxide (and oxygen)

Notice that these networks are arranged hierarchically. This reflects the hierarchical nature of anesthetic decision making: global decisions (e.g., whether to use a general or regional anesthetic technique) involve subdecisions (e.g., whether or not to intubate), which involve subdecisions (e.g., which intubation technique to use), which may involve further subdecisions (e.g., which muscle relaxant to use to facilitate intubation).

The bottommost decisions involve specific anesthetic agents or techniques. These are called "elemental" arcs since no further subdecisions are required to specify exactly how that technique is to be performed. The higher level decisions, however, do require further subdecisions and are therefore called "nonelemental" arcs. As shown in Figure 3.5, these subdecisions are spelled out by other ATN networks, lower in the hierarchy.

In using the ATN to explore alternative anesthetic approaches, ATTENDING begins at the topmost network (ANES) and starts exploring different paths. Each time a "nonelemental" arc is chosen, exploration in that network is temporarily suspended while the indicated subnetwork is explored (which may in turn involve the exploration of further subnetworks). Finally, when exploration of the lower network is completed, exploration of the higher network resumes.

In performing this analysis, ATTENDING may explore alternative paths at several levels. For instance, Figure 3.6 expands upon the simple example of Figure 3.4. Here, the physician has suggested maintenance of general anesthesia using halothane and this involves some risk in the patient described. ATTENDING may therefore explore alternative approaches at three different levels.

1. As shown in Figure 3.6, ATTENDING first looks at the lowest level to see whether an alternative inhalational agent might involve less risk.
2. If all inhalational agents involve risk, however, then ATTENDING can explore alternatives at an intermediate level, to see whether a balanced (light) anesthetic technique might involve less risk than an inhalational technique.
3. Finally, if there are risks involved in all inhalational and balanced anesthetic techniques, then ATTENDING can explore more global alternatives, such as regional anesthesia.

In this way, using the hierarchical ATN formalism, ATTENDING first searches for alternatives that involve the smallest change from the proposed plan, and only if these involve risk need it explore more global changes.

3.5.1.1. Internal Representation of ATTENDING's ATNs. This section shows how ATTENDING's anesthetic management ATNs are represented internally as LISP data structures. Several sample networks are shown below.

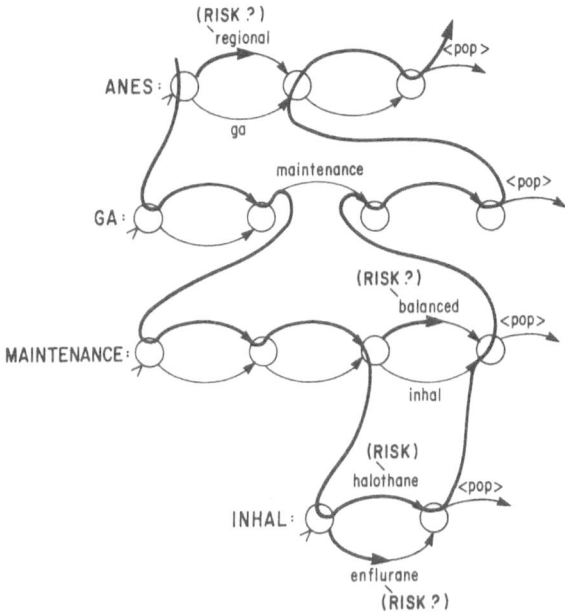

Figure 3.6. ATTENDING may explore alternative paths at any level.

```
(ANES  (GA AN1 TGA)
        (REGIONAL AN1 TREG))
(AN1  (PREMED AN2 TPREM)
        (NOPREMED AN2 TNOPREMEDS))
(AN2  (POP T T))

(GA  (INTUBATION G2 TINTUB))
(G2  (MAINTENANCE G3 TMAINT))
(G3  (INDUCTION G4 TINDUCT))
(G4  (POP T T))

(INTUBATION  (AWAKE INT1 TAWAKE)
             (NONE INT1 TNOINTUB)
             (MASKCRICOID INT1 TMASKCRICOID)
             (RAPIDSEQ INT1 TRAPIDSEQ)
             (CONVENTIONAL INT1 TNORMINT))
(INT1  (POP T T))
```

In this internal form, each state starts with a state name (ANES, AN1, AN2, GA, etc.) and is followed by a set of arcs, each of which has three components: (1) the name of an anesthetic technique associated with the arc (e.g., INTUBA-TION), (2) the name of the arc's "destination state" (e.g., G2), (3) a name unique to the arc (e.g., TINTUB).

The unique name serves two functions. As described in Section 3.6, it is used to indicate which risks and benefits apply to the arc. Also, for some arcs, this name also indicates an "action routine," a LISP program that tests contextual information and inactivates the arc in certain circumstances. For instance, the spinal anesthesia arc would be inactivated for operations on the head and neck, for which spinal anesthesia is impossible. (Action routines are part of the ATN formalism, but are not used extensively in ATTENDING.)

3.5.1.2. How ATTENDING Uses the ATN. ATTENDING analyzes an anesthetic plan in three stages, as shown in Figure 3.7. First, ATTENDING converts the plan into a tree-structured form, which reflects the hierarchical nature of the decisions involved. (The construction of this tree is accomplished by a "top-down" processing of the ATN networks.)

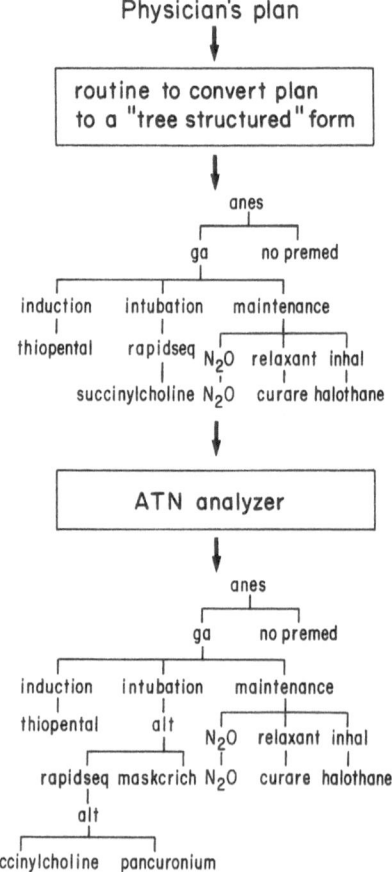

Figure 3.7. The stages of ATTENDING's ATN analysis.

Once this tree is constructed it is input to an "ATN analyzer," which uses the ATN to explore alternative approaches. The ATN analyzer produces an "augmented" tree, which includes the original plan and may include alternatives at any level. This augmented tree forms the basis for the prose critique.

The exploration of the ATN proceeds as follows:

1. Starting with the initial state of a network (state A), ATTENDING first sees which arc (arc a) the physician has proposed, in the plan tree. (ATTENDING also evaluates the risks associated with arc a, as described in Section 3.6.)
2. ATTENDING then examines every other arc leaving state A. For each such arc (arc b), ATTENDING evaluates the risks associated with arc b.
3. If arc b involves more risk than arc a, or than some other arc from state A, then arc b is discarded from further analysis. Otherwise, arc b is remembered for inclusion in the augmented tree and the critique.
4. Processing then continues at the next state of the network.

These four steps give a rough overview of how ATTENDING's ATN analyzer operates. Two further features are: (1) If the proposed plan involves *no risk*, then only alternative plans with benefits are remembered. (There is no point in mentioning alternative choices when the physician's choice is satisfactory.) (2) As discussed in Section 3.6, "contextual preference rules" may be used to help pare down the number of alternatives.

The result of this analysis is a set of alternatives that involve either less risk or roughly equivalent but different risks compared to the proposed approach. These alternatives are then discussed in the critique.

The exact mechanism used by ATTENDING for evaluating and comparing risks is discussed in Section 3.6. The remainder of this section discusses the ATN formalism from several perspectives to help the reader understand its advantages and its limitations as a model for medical management. (ATNs are *not* used to structure decision making in the other systems described in this book. The remainder of this section explains in part why.)

3.5.2.1. Advantages of the ATN. There are a number of potential advantages of the ATN formalism as a model of anesthetic decision making. First of all, as discussed above it captures the *hierarchical* nature of anesthetic decisions. On the other hand, the three other systems discussed in this book do not use an explicitly hierarchical model. Instead, they handle the hierarchical nature of their domains merely by discussing high-level decisions before discussing subdecisions, etc. Thus, in these other systems, the hierarchical structure is handled solely by the order in which critiquing comments are made.

Another possible advantage of the ATN formalism can be seen by comparing it to a decision tree, a formalism widely used in clinical decision analysis. In a decision tree, decision branches continually diverge. In contrast, in an ATN, paths that initially diverge may later rejoin. For instance, in ATTENDING's maintenance network, one first decides whether to use nitrous oxide, then whether to use a muscle relaxant, and finally whether to use a balanced or an inhalational technique. Since ATN arcs may rejoin, these three decisions are modeled very naturally. In a decision tree, successive decisions would have to be spelled out redundantly in divergent branches of the tree. As a result, the ATN offers a more economical model for capturing the decision-making process.

As discussed in Chapter 9, the decision-making process in anesthesiology has a somewhat different character than in the other domains explored in this book. There tend to be more choices at each decision point, each with a number of possible risks and benefits. As a result, there is a need for a model like the ATN to allow ATTENDING to explore alternatives dynamically and compare risk. This dynamic exploration of alternatives was not required in the other domains, which has allowed their internal design to be considerably simpler than ATTENDING's.

3.5.2.2. Local vs. Global Plan Analysis. A significant limitation of the ATN model as used in ATTENDING is that the various subdecisions are assumed to be independent of one another. In other words, a subdecision made in one part of the plan is assumed not to affect decisions made elsewhere.

In contrast, several nonmedical AI research projects have explored the problems of planning in domains where subplans do interact (Sacerdoti 1977; Stefik 1981). In such a domain, decisions cannot be analyzed independently of other parts of the plan.

In ATTENDING's domain, only one situation was found where subplans did interact: if a patient with an airway problem is intubated awake, certain problems disappear that might otherwise exist during induction. Rather than develop a general mechanism to let ATTENDING handle this one situation, it is handled as a special case in an ad hoc fashion.

3.6. Evaluating Risk and Risk Tradeoffs

The previous section describes how ATTENDING is designed to explore alternative approaches to a patient's anesthetic management. In addition, the system must compare the various risks involved in the different techniques.

3.6.1. Problem Management Frames

The knowledge which ATTENDING uses to assess risk is contained in "problem management frames" (PM frames). Each underlying medical problem has an associated PM frame which outlines the anesthetic implications of that problem: the various possible risks and benefits that the problem implies for different anesthetic techniques.

An example PM frame is shown below, outlining the anesthetic implications of asthma:

```
(DEFFRAME 'ASTHMA '(
PRINCIPLES (
  (PREFER DEEP_ANESTHESIA DANGER BRONCHOSPASM)
  (PREFER BRONCHODILATORS DANGER BRONCHOSPASM)
  (AVOID BRONCHOSPASTICS DANGER BRONCHOSPASM)
  )
RISKS (
  (ARC TRAPIDSEQ RISKVAL LOWR DESC (BRONCHOSPASM)
    WHENSEVERE T)
  (ARC TINHAL RISKVAL LOWB DESC (PREVENTING BRONCHOSPASM
    BY KEEPING THE ANESTHETIC LEVEL DEEP) WHENSEVERE T)
  (ARC THALO RISKVAL LOWB DESC (HELPING PREVENT
    BRONCHOSPASM SINCE HALOTHANE IS A BRONCHODILATOR)
    WHENSEVERE T)
  (ARC TKETAMINE RISKVAL MODB DESC (SUPPRESSING
    BRONCHOSPASM SINCE KETAMINE IS A BRONCHODILATOR)
    WHENSEVERE T)
  (ARC TDTC RISKVAL LOWR DESC (BRONCHOSPASM) MECH
    HISTAMINE MECHDESC (TRIGGERED HISTAMINE RELEASE)
    WHENSEVERE T)
  (ARC TMORPH RISKVAL LOWR DESC (BRONCHOSPASM) MECH
    HISTAMINE MECHDESC (DUE TO HISTAMINE RELEASE)
    WHENSEVERE T)
  (ARC TMORPHPM RISKVAL LOWR DESC (BRONCHOSPASM) MECH
    HISTAMINE MECHDESC (DUE TO HISTAMINE RELEASE)
    WHENSEVERE T)
  (ARC TETH RISKVAL LOWR DESC (INITIAL BRONCHOSPASM DUE TO
    AIRWAY IRRITABILITY) WHENSEVERE T)
  (ARC TTPL RISKVAL LOWR DESC (BRONCHOSPASM) WHENSEVERE
    T)
  (ARC TAWAKE RISKVAL MODR DESC (BRONCHOSPASM)
    WHENSEVERE T)
  (ARC TBALANCED RISKVAL LOWR DESC (POSSIBLE
    BRONCHOSPASM DUE TO LIGHT ANESTHESIA))
  (ARC TNOSED NOTCONTEXT (ACUTE_TRAUMA) RISKVAL MODR
    WHENSEVERE T DESC (PROBLEMS WITH THE PATIENT
    APOST_S ASTHMA)
  )))
```

This PM frame has two components. First, it indicates three underlying anesthetic management principles implied by asthma. Second and most importantly, it lists the various risks and benefits of different anesthetic techniques for a patient with asthma. For instance, the first risk description is:

```
(ARC TRAPIDSEQ RISKVAL LOWR DESC (BRONCHOSPASM)
  WHENSEVERE T)
```

This risk description says, in effect, "Rapid-sequence intubation has a low risk of provoking bronchospasm (in an asthmatic patient), especially when the asthma is severe." The risk description includes several components:

1. A pointer to the ATN arc (i.e., the anesthetic technique) to which the risk applies, in this case, TRAPIDSEQ
2. A rough estimate of the magnitude of the risk, LOWR (low risk), as discussed in Section 3.6.3
3. Additional information used in discussing the risk in the prose critique

These risks and benefits may apply to elemental arcs (individual anesthetic agents or techniques) or to nonelemental arcs (e.g., inhalational anesthesia, intubation, general anesthesia).

Figure 3.8 illustrates schematically how ATTENDING's PM frames function. For each underlying medical problem, a PM frame outlines the risks and benefits which that problem implies for each arc of ATTENDING's ATN. These risks are then used when exploring the ATN, as described previously.

3.6.2. Conventional Approaches to Risk Analysis

Before describing how ATTENDING handles risk, it is useful to describe how risk analysis is conventionally approached, for instance, in clinical decision analysis (Weinstein et al. 1980). Risk analysis is usually handled statistically. Each decision is broken down into all its possible outcomes (good and bad results). Each such outcome (i) is then assigned:

1. a numerical "likelihood" (L_i)
2. a numerical "value" (V_i), which represents the cost or benefit of the outcome along a numerical scale, such as dollars or "quality of life units"

Once a decision is broken down in this fashion, the "expected value" of making that decision can be computed: the sum of each outcome's value multiplied by its likelihood. The expected value of different decisions can then be compared to see which is most desirable.

This statistical approach to risk analysis is computationally straightforward. When applied to complex problems in the real world, however, the approach involves significant practical problems. (1) The various numbers required may

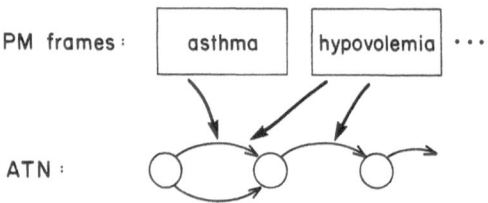

Figure 3.8. The risks and benefits in ATTENDING's PM frames are associated with ATN arcs.

be very difficult to determine. (2) Any numbers obtained may be the product of several underlying assumptions, upon which the entire analysis is therefore based.

1. Likelihood. Obtaining the likelihood of all possible outcomes may be difficult, if not impossible. For one thing, different studies published in the clinical literature may contain significantly different results and, in any case, may have different study designs, different patient populations, etc. Furthermore, it is impossible to obtain statistics for *poor* decisions. The only way to obtain such statistics would be to deliberately manage a group of patients poorly, and carefully document all the bad results that occurred: a clearly impractical project.
2. Outcome values. It is similarly difficult to estimate the numeric values to be associated with different outcomes. Indeed, a significant portion of economic cost/benefit literature addresses the difficulties of estimating real-world costs and benefits. How, for example, does one assign a number to the loss of a limb, of an eye, or of a life? How does that number change if the patient is already chronically ill? How are numbers assigned to such complications as hypotension, bronchospasm, etc.?

In addition, of course, the gathering of these statistics would be a major, even prohibitive task. In the face of all these difficulties, it is reasonable to ask whether a statistical analysis is required for a system like ATTENDING.

Certainly physicians deal with risk every day with only a rough feel for the magnitude of the risks involved. Instead of formal statistical analysis, they take a "heuristic" (approximate) approach to risk. This is also the approach taken by ATTENDING. The remainder of this section describes the heuristic approach to risk analysis, which has been implemented to allow ATTENDING to evaluate risk and to assess risk tradeoffs.

3.6.3. A Heuristic Approach to Risk Analysis

As discussed above, ATTENDING does not take a statistical approach to risk analysis but has adopted a heuristic approach. This heuristic approach allows ATTENDING to deal with issues of risk without reducing risk to numbers which are only approximations of reality in any case. The physician is then allowed to use his judgment in making the final subjective assessment of any particular risk tradeoffs in his patient.

The heuristic approach to risk analysis is one of the interesting features of ATTENDING's design. It serves to highlight the centrality of risk in medical management and also demonstrates how risk may be handled by an expert consultation system.

To allow it to compare risks and to discuss tradeoffs in an intelligent fashion, ATTENDING adopts a heuristic approach with three components:

1. First, ATTENDING uses *rough estimates of risk magnitude* to allow it to perform a "broadbrush" comparison of different risks. This broadbrush comparison allows it to discard management choices that are clearly inferior, and to recognize choices that are clearly superior.
2. Next, ATTENDING uses *contextual preference rules* to allow it to focus selectively on approaches that might be particularly appropriate for certain patients.
3. Finally, ATTENDING uses *pragmatic features* associated with the various risks to let it couch its prose discussion of the risks in appropriate terms.

Figure 3.9 illustrates schematically how these steps allow ATTENDING to focus on a subset of possible approaches for its final discussion with the physician. This process is applied at each level of decision and subdecision in ATTENDING's analysis of the user's plan.

3.6.3.1. Rough Risk Magnitude. When a human contemplates a decision, a number of clearly inferior choices may not even come to mind. These are presumably weeded out at an unconscious level. Of course, if a computer is to do this, the required knowledge must be explicitly built in. ATTENDING achieves this by assigning each risk a "rough risk magnitude": LOW, MODERATE, HIGH, or EXTREME.

Formally, such a risk magnitude corresponds to the sum of the likelihood times the value for all possible outcomes implied by the risk. Clinically, these rough risk magnitudes seem to be quite easy to assign by practicing anesthesiologists. Clearly, it is much easier to estimate this value than to compile a statistical approximation.

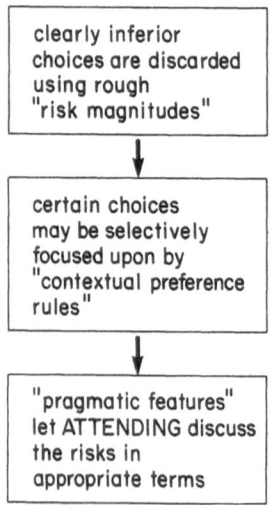

Figure 3.9. The three steps in ATTENDING's heuristic risk analysis.

Examples of rough risk magnitudes are:

1. LOW. An example of a LOW risk is the use of the inhalational anesthetic enflurane in a patient with renal failure. Although there are theoretical reasons to avoid this, there are no documented cases where this has caused problems.
2. MODERATE. An example of a MODERATE risk is using morphine in a patient with asthma. Depending on the severity of the patient's asthma, morphine may provoke bronchospasm because it releases histamine.
3. HIGH. An example of a HIGH risk is using succinylcholine in a patient with a penetrating eye wound. This has a reasonable chance of resulting in the loss of the eye.
4. EXTREME. An example of an EXTREME risk is not intubating a patient with a full stomach; this could well result in aspiration of stomach contents into the lungs which is often fatal.

In addition to these rough estimates of risks, ATTENDING uses similar estimates for benefits. In its internal analysis, ATTENDING treats benefits as "negative risks." Most benefits in anesthetic management fall into the LOW to MODERATE range.

In its internal evaluation of these risk magnitudes, ATTENDING operates as follows:

1. If an alternative under consideration involves several risks, the overall risk assigned to that decision is the highest of the individual risk magnitudes: i.e., LOW + LOW = LOW, LOW + MODERATE = MODERATE, LOW + HIGH + HIGH = HIGH.
2. When comparing several different choices, ATTENDING remembers only those whose rough risk magnitudes are lowest. All choices with higher risk magnitude are discarded from further analysis. (The only exception to this rule is that the physician's choice is always remembered since it serves as a standard for comparison.) As a result, after comparing a set of alternatives, ATTENDING is left with the physician's choice plus one or more alternatives that either involve less risk or involve roughly equivalent, but different, risks.

In this way, using rough risk magnitudes, ATTENDING is able to focus its attention on these reasonable approaches in its later analysis and in its prose critique.

3.6.3.2. Contextual Preference Rules. Once the analysis of rough risk magnitudes has selectively focused attention on a subset of possible choices, a second level of analysis may then be used, depending on the particular patient and the plan.

This second level of analysis involves the use of "contextual preference rules" (CP rules). These CP rules allow ATTENDING to express a preference between two choices of the same rough risk magnitude. A CP rule has four components indicating:

1. In the presence of a particular *medical problem*
2. *Approach A* may be preferred
3. To *approach B*
4. For a *specified reason*

The following example shows how one CP rule is expressed in ATTENDING's knowledge base:

(MOREPREF TMASKCRICOID LESSPREF TRAPIDSEQ CONTEXT CAD
   WHY (COMMA IF SEVERE COMMA SINCE IT ALLOWS MORE GRADUAL
   INDUCTION WITH LESS DANGER OF SUDDEN CARDIAC
   DECOMPENSATION))

This rule states that (1) in the presence of coronary artery disease (CAD), (2) intubation after mask induction with cricoid pressure (TMASKCRICOID), (3) may be preferred to rapid-sequence intubation (TRAPIDSEQ), (4) "since it allows more gradual induction with less danger of sudden cardiac decompensation."

CP rules are used in two ways in ATTENDING's analysis of the physician's plan. First, if a rule applies and the physician has *not* suggested the less preferred approach, then that choice is discarded from further analysis. Alternatively, if the rule applies and the physician *has* suggested the less preferred approach, then the CP rule allows ATTENDING to indicate in its critique that a preferred approach may exist. The CP rule also allows ATTENDING to give a *reason* for its preference. For example, the following partial critique illustrates the application of the CP rule shown above.

> Intubation of this patient is definitely appropriate because of a full stomach. Rapid-sequence intubation is a reasonable approach. This might, however, have the risk of cardiac compromise. An alternative approach to rapid-sequence intubation would be intubation after mask induction with cricoid pressure. This has the advantage of helping avoid hypotension. Intubation after mask induction with cricoid pressure might well be preferable for a patient with coronary artery disease, if severe because it allows more gradual induction with less danger of sudden cardiac decompensation.

In summary, then, ATTENDING has two mechanisms that allow it to selectively focus its attention on a relevant subset of possible choices: (1) the analysis of rough risk magnitudes, and (2) the contextual preference rules. The first of these mechanisms is a general heuristic applied to all risks. The second mechanism is a selective one, tailored to particular clinical situations.

It is worth pointing out that, in designing ATTENDING, we might have tried to use only the risk magnitude analysis. In particular, we might have tried to subdivide the risk magnitude categories into finer gradations to accommodate distinctions currently captured by CP rules. This approach was not adopted for several reasons:

1. By keeping the risk magnitude analysis simple, that analysis is easily understood, and it is quite easy to assign risks appropriately. If instead we allowed

a larger number of risk categories, the distinctions would become harder to define, and the simplicity of the approach would be lost.

2. By using CP rules, the risk magnitude analysis remains simple and is augmented in an explicitly stated way. Each CP rule states precisely where and when it applied, and also states the reason for its preference.

Thus, CP rules refine the risk magnitude analysis in an explicit fashion, and also allow ATTENDING to state its rationale for any finer preferences that it advocates.

3.6.3.3. Pragmatic Features. The third component of ATTENDING's heuristic approach to risk analysis involves "pragmatic features" associated with many of the risks to allow the system to couch its discussion of those risks in appropriate terms. Examples of pragmatic features used by ATTENDING in discussing risks include:

1. IMPLICIT. Some risks are so obvious to the physician that any discussion of them sounds artificial. An example might be the risk of aspiration when performing a rapid-sequence intubation. These risks are considered internally when the plan is analyzed but need not be discussed in the prose critique.
2. THEORETICAL. Some risks are really only theoretical and of little practical importance. When mentioning such risks, ATTENDING couches its discussion in such phrases as "the theoretical risk of _____."
3. REMOTE. Other risks, while real, are very remote. ATTENDING couches its discussion of such risk with such phrases as "the conceivable risk of _____."

If these pragmatic features are not used, when ATTENDING discusses the risks of a set of alternatives, it sounds like a schoolchild reciting facts it has memorized but does not understand. For example:

> Intubation of this patient is definitely appropriate because of a full stomach. Rapid-sequence intubation is a reasonable approach. This, however, might have two possible risks. First, there is the risk of bronchospasm. Second, there is the risk of aspiration. An alternative approach to rapid-sequence intubation would be intubation after mask induction with cricoid pressure. This has the risk of aspiration.

In contrast, when the pragmatic features are used, the discussion sounds more natural to the anesthetist (who usually does not even notice that the phrasing of different risks is being couched in somewhat different terms).

> Intubation of this patient is definitely appropriate because of a full stomach. Rapid sequence intubation is a reasonable approach. This, however, might have the conceivable risk of bronchospasm. An alternative approach to rapid-sequence intubation would be intubation after mask induction with cricoid pressure.

3.6.3.4. Summary: Heuristic Risk Analysis. Section 3.6 has discussed ATTENDING's heuristic approach to risk analysis. This approach was developed to let ATTENDING assess risk without a massive compilation of statistics. Instead, the approach uses a spectrum of different types of knowledge designed to let ATTENDING heuristically evaluate risk tradeoffs internally and discuss them intelligently in its critique.

3.7. PROSENET: Flexible Generation of English Prose

Once ATTENDING has explored alternative approaches to a patient's anesthetic management and evaluated any risks and risk tradeoffs, the final step is the creation of a prose critique. This is accomplished using PROSENET, an approach developed to facilitate the generation of polished prose.

PROSENET was initially developed for use in the ATTENDING system but proved to be a very flexible tool. It was therefore used in implementing all the critiquing systems described in this book and was also incorporated into the ESSENTIAL-ATTENDING system-building system. Since PROSENET itself is described in Chapter 7, which describes ESSENTIAL-ATTENDING, it will not be described in detail here.

3.8. ATTENDING: Limitations

The preceding sections have described the internal design of the ATTENDING system. It is important to emphasize that ATTENDING is not at present a practical consultation system, but rather a research prototype developed to explore basic design issues. This section discusses some of the ATTENDING's current system limitations.

ATTENDING's knowledge of anesthetic management is limited in several ways. Two limitations involve the two knowledge bases discusses earlier in this chapter: (1) its knowledge of anesthetic management techniques, and (2) its knowledge of the anesthetic implications of underlying medical problems.

1. Anesthetic techniques. ATTENDING currently includes most of the commonly used anesthetic agents and techniques. There are a number of other techniques that are not included, however, especially some of the ones less commonly used. Also, different agents and techniques may be combined. ATTENDING is currently designed to discuss only "pure" techniques.
2. Anesthetic implications of underlying diseases. As discussed previously, ATTENDING currently knows the anesthetic implications of only 25 underlying medical problems. This is its major knowledge limitation. For comprehensive practical use, the knowledge of considerably more diseases would be required.

In addition, ATTENDING has other more subtle limitations. For instance, its knowledge of how different medical problems might interact is limited. Also, as discussed previously, there is a great deal of more detailed information that a critiquing system may usefully provide. In fact, the next chapter discusses how the HT-ATTENDING system, which critiques hypertension management, includes much more detailed descriptive information.

3.9. ATTENDING: Summary

The implementation of the ATTENDING system was our first step in exploring the critiquing approach. From this experience, we learned a great deal as to what capabilities are easy to implement, what capabilities are difficult, and what different types of advice a critiquing system might offer. As a result of this experience, the three other systems described in this book have adopted a considerably different structure.

1. They do *not* use the ATN to model medical decision making.
2. They do *not* perform an explicit dynamic analysis of risk.
3. As mentioned above, they allow much more descriptive material to be included in the critique.

As a result, the systems discussed in the next three chapters represent a significant departure from ATTENDING's design. In part, this is a reflection of different characteristics of their domains. It also reflects, however, our experience implementing ATTENDING, and a resulting desire (1) to simplify the internal system design, while at the same time (2) to allow a more clinically robust critiquing output.

Chapter 4

HT-ATTENDING: Essential Hypertension*

The HT-ATTENDING system extends the exploration of the critiquing approach to the domain of essential hypertension. This domain is interesting for several reasons. First, essential hypertension is a widespread medical problem, for which treatment is most important. Second, there is a complex spectrum of treatment choices, with new drugs frequently introduced. As a result, the field is in constant flux and it is important that a physician have access to accurate, up-to-date information. Essential hypertension is also a good domain because it is familiar to most physicians. In contrast, anesthesiology is quite esoteric. Most physicians know little, if anything, about anesthetic management.

HT-ATTENDING's design differs from ATTENDING's in two major ways: (1) in the "depth" of the internal analysis that the two systems perform, and (2) in the breadth of the issues dealt with in the prose discussion. Paradoxically, ATTENDING performs a "deeper" internal analysis, but HT-ATTENDING goes into much more clinical detail in its prose critique.

ATTENDING's analysis is "deeper" because it uses explicit knowledge about the various risks involved in anesthetic management, and it manipulates and compares these risks explicitly. ATTENDING's critique, however, is a fairly superficial discussion of alternative approaches and of the various risks involved. Although ATTENDING's critique gives a coherent overview of these risks, little detailed descriptive material is included.

In contrast, HT-ATTENDING's design allows much more flexibility to produce detailed descriptive prose. The system's "author" can include in the system's prose whatever material and whatever nuances he considers important. Thus, HT-ATTENDING's design differs from ATTENDING's in two fundamental ways:

---

*This chapter is adapted from Miller, P.L. and Black, H.R.: Medical plan-analysis by computer: Critiquing the pharmacologic management of essential hypertension. Computers and Biomedical Res. 17:38–54, 1984. Copyright 1984 by Academic Press, Inc. Reprinted by permission of the Publisher.

1. HT-ATTENDING does not explicitly manipulate and compare risks and benefits in its internal analysis. As a result, it does not tailor the set of alternatives discussed as flexibly as ATTENDING does. Instead, HT-ATTENDING discusses any risks or benefits of the physician's plan, but then discusses alternatives only in fairly general terms. As discussed in Chapter 9, certain characteristics of the hypertension domain make this more "reactive" critiquing design possible.

2. HT-ATTENDING's prose critique is assembled from larger "chunks" of prose than ATTENDING's. As a result, HT-ATTENDING's author has much more flexibility to say in detail what he feels the user should hear about a particular issue.

A corollary of HT-ATTENDING's design, however, is that the system itself does not really "understand" the prose that it produces as deeply as ATTENDING did. (Even ATTENDING's "understanding" of its advice, of course, is very limited.) Whereas ATTENDING's critique is based on a comprehensive analysis of the underlying risks involved, HT-ATTENDING's analysis consists of "preprogrammed" reactions to specified sets of conditions with appropriate prose comments.

In one sense, HT-ATTENDING's design may be seen as a step back from a truly "expert" system, since the system itself understands its domain less deeply. The approach does, however, let HT-ATTENDING give very robust advice to its user, even if its own "comprehension" of that advice is less deep.

Indeed, HT-ATTENDING's design has led to the realization, discussed later in this chapter, that a critiquing system can be modeled as an "interactive paper." The domain expert is the author of this paper. The user physician is the reader. The paper itself adapts its contents to a particular approach to a particular patient's care. In fact, HT-ATTENDING was developed in collaboration with Dr. Henry R. Black, a hypertension specialist, and is modeled closely after a paper he wrote on hypertension management.

The remainder of this chapter describes the initial implementation of HT-ATTENDING and the various issues that arose in its design. This initial system, developed in 1983, has since been converted to the ESSENTIAL-ATTENDING format (see Chapter 7) and is now being updated to reflect current (1985) thinking in the field.

To use HT-ATTENDING, a physician first answers a modest number of questions concerning a patient, including current blood pressure, sex, age, underlying medical problems, and concurrent medications. The physician also describes the patient's present antihypertensive regimen, together with a proposed change in that regimen.

HT-ATTENDING then produces a prose critique discussing the risks and benefits of the proposed approach and a discussion of other approaches that may be either reasonable or preferred. The purpose of the critique is (1) to assist the physician in managing a hypertensive patient, (2) to help him avoid inadvertent management errors, (3) to inform him about relevant new drugs and treatment modalities, and (4) to mention topical issues, all in the context of discussing an approach to a particular patient's care.

4.1. Examples

This section shows four examples of HT-ATTENDING critiquing proposed approaches to different patients' antihypertensive management. To use the system, a physician first answers questions describing his patient, as illustrated below. (In this sample session, the physician's responses are underlined.)

Please describe your patient.

age ***48
sex: 1. male   2. female
type one ***1
complicance assessment: 1. good   2. poor   3. unknown
type one ***1
systolic pressure ***160
diastolic pressure ***104

Are any of the following concurrent conditions present?
   1. asthma
   2. cardiac conduction delay
   3. congestive heart failure
   4. cerebrovascular disease
   5. coronary artery disease
   6. depression
   7. diabetes mellitus
   8. gout
   9. hyperthyroid
   10. hypothyroid
   11. liver disease
   12. obesity
   13. peptic ulcer disease
   14. peripheral vascular disease
   15. renal insufficiency
   16. syncope
   17. concurrent digitalis therapy
   18. concurrent phenothiazine therapy
   19. concurrent tricyclic therapy

type 0, or any numbers that apply ***8

Here the physician has described a 48-year-old man with a history of gout, whose blood pressure is 160/104. Next the current and proposed antihypertensive medications, including dosages, are indicated. The process of entering all this information into the computer takes a minute or two.

HT-ATTENDING then produces its prose critique. Four such critiques are shown below. Following each, a number of illustrative features of that critique are briefly discussed. A more detailed discussion follows in the remainder of the chapter.

Example 1. An otherwise healthy 48-old man with gout. BP 160/104.

current regimen: none
proposed regimen: hydrochlorothiazide 50 mg bid

In the United States, diuretics are recommended by most authorities as initial or step one therapy. Since the initiation of diuretic therapy may precipitate a gouty attack, you may wish to start prophylactic treatment when initiating a diuretic in a patient with gout.

A thiazide diuretic or similar acting agent is the drug of choice for most patients. Hydrochlorothiazide is a commonly used thiazide diuretic. Chlorthalidone or metolazone are alternative agents worth considering since they are long acting and can be given once a day. We usually start treatment with a lower dose than the dose you propose. We now recommend starting treatment with hydrochlorothiazide 25 mg qd or bid. Recent work by Materson [1] has shown that low doses of diuretics may be as effective as higher doses yet cause less hypokalemia.

When initiating antihypertensive medications, the clinician must first satisfy himself that (1) the possibility of an underlying etiology has been adequately ruled out, (2) nonpharmacologic therapy is insufficient, e.g., salt restriction and diet, and (3) the patient is not taking other medications contributing to the hypertension, such as birth control pills, steroids, nonsteroidal antiinflammatories, or appetite suppressants.

This example demonstrates several features of HT-ATTENDING's design. The section on diuretics shows the system discussing both alternative agents and alternative doses. The system will also discuss dose frequency when appropriate. The section on diuretics also shows the system documenting one of its recommendations with a reference. This citation (Materson et al. 1978) is available online upon request.

The final paragraph is significant in that it is not really part of the critique. HT-ATTENDING concentrates its critique solely on the *pharmacologic management* of essential hypertension. It does not attempt, for instance, to ask about, or critique, the adequacy of the physician's workup of the patient for underlying etiologic factors. Nevertheless the system includes this paragraph in its output as a reminder to the physician whenever a patient is started on antihypertensive drugs.

Example 2. A 53-year-old man with a history of asthma and mild renal insufficiency. BP 156/102.

current regimen: hydrochlorothiazide 100 mg bid
proposed regimen: hydrochlorothiazide 100 mg bid, propranolol 40 mg bid

In the United States, diuretics are recommended by most authorities as initial or step one therapy. A thiazide diuretic or similar acting agent is the drug of choice for most patients. However, if this patient has significant renal insufficiency (glomerular filtration rate less than 50 ml per minute), than a loop diuretic like furosemide would generally be considered most appropriate.

If you do use a thiazide diuretic, chlorthalidone or metolazone are alternative agents worth considering since they are long acting and can be given once a day.

We would not use as high a dose as you propose. Once a daily dose of hydrochlorothiazide 50 mg or its equivalent is reached and given an adequate trial, most authorities now feel that no further increase in diuretic dosage is indicated and that other drugs should be added. Some patients, however, perhaps as many as 15 to 20 percent of the group that has not responded to this dose, may come under control when the dose is doubled. These patients would thus be spared the additional problems of other agents.

A beta blocker is, in general, our preferred step two agent. In a patient with asthma, however, we would recommend avoiding beta blockers entirely in favor of alternatives such as central sympathetic blockers or possibly captopril. Although it is clear that patients with asthma should not generally be given beta blockers, should it be necessary to do so, cardioselective agents (metoprolol and atenolol) may be better tolerated and asthmatic attacks precipitated by beta blockage may be more easily reversed.

Before progressing to a step two agent, it may be worth making sure the patient has been taking medications faithfully, and adequately restricting sodium intake.

Here again in the section on diuretics, HT-ATTENDING discusses alternative drugs that might be appropriate (in this example, furosemide is included since the patient has renal insufficiency), as well as discussing the proposed dose. Also, the system is seen responding to two possible management errors: the use of a beta blocker in a patient with asthma, and the use of a thiazide diuretic in the presence of renal insufficiency.

The last paragraph here again is not really an integral part of the critique but is included as a reminder to the physician whenever step two treatment is initiated.

Example 3. A 59-year-old woman with a history of congestive heart failure. BP 162/102.

current regimen: chlorthalidone 50 mg qd, propranolol 160 mg bid
proposed regimen: chlorthalidone 50 mg qd, propranolol 160 mg bid, guanethidine
                20 mg qd

In the United States, diuretics are recommended by most authorities as initial or step one therapy. A thiazide diuretic or similar acting agent is the drug of choice for most patients. Chlorthalidone has the advantage that it is long acting and can be given once a day.

A beta blocker is, in general, our preferred step two agent. Clearly, in a patient with decompensated congestive heart failure, however, all beta blockers should be avoided in favor of alternatives, such as central sympathetic blockers or possibly captopril. On the other hand, in certain patients with well-compensated congestive heart failure, a beta blocker may be useful. Here pindolol, which has intrinsic sympathomimetic activity (ISA) may well be the beta blocker of choice.

Guanethidine should be reserved for patients who have not responded to other agents since it is often poorly tolerated and difficult to administer. Thus for this patient, we would suggest first trying a step three drug, such as hydralazine, prazosin, or possibly captopril before using guanethidine, which we reserve for use as a step four agent. When treatment with guanethidine is begun, we recommend that all other treatments except diuretics be stopped. If you do use guanethidine, we would suggest starting treatment with a lower dose than you propose. We recommend that guanthidine be given once a day starting with a dose of 10 mg and increased slowly until the desired therapeutic effect is achieved or intolerable toxicity occurs.

As in example 2, example 3 shows a patient receiving a beta blocker despite a relative contraindication. Since the contraindication with congestive heart failure is not as strong as with asthma, the tone of the critique is correspondingly milder. If, after reading the critique, the physician would like to know more about certain topics (such as pindolol and ISA), he can ask to see one of several discussions which HT-ATTENDING stores online (for example, "The Six Beta Blockers Presently Available in the United States").

This example illustrates the fact that HT-ATTENDING takes a stepped-care approach to hypertension management. In the third paragraph, for instance, the system reacts to the initiation of the potent drug guanethidine before milder alternatives are tried and found to be inadequate. HT-ATTENDING's preference, which it shares with most authorities, is to reserve guanethidine for use as a step four drug. The rationale for allowing a computer system to have its own "practice preferences" is discussed later.

Example 4. A 40-year-old otherwise healthy man. BP 150/100.

current regimen: none
proposed regimen: prazosin 2 mg bid

Prazosin has been used by some as sole therapy, but is not as effective as a diuretic or a beta blocker for this purpose. In the United States, diuretics are recommended by most authorities as initial or step one therapy. If you do use prazosin, the starting dose you propose is higher than we would suggest. We usually recommend a starting dose of prazosin of 1 mg bid and a maximum dose of 10 mg bid. The dose may be increased every 2–3 days. While the side effects of prazosin are generally mild, the clinician must be careful to avoid the "first dose phenomenon" of syncope or presyncope commonly seen with this agent when the initial dose is too large. The first dose of the drug should be given at night and the patient cautioned against too rapid change to the upright posture. This problem is rarely seen after the initial dose, although it has been noted when the dose is increased.

When initiating antihypertensive medications, the clinician must first satisfy himself that (1) the possibility of an underlying etiology has been adequately ruled out, (2) nonpharmacologic therapy is insufficient, e.g., salt restriction and diet, and (3) the patient is not taking other medications contributing to the hypertension, such as birth control pills, steroids, nonsteroidal antiinflammatories, or appetite suppressants.

Here HT-ATTENDING again expresses certain of its practice preferences. Also, since drug therapy is being initiated, paragraph 2 is again produced as a reminder to the physician.

These examples serve to illustrate several reasons why the pharmacologic management of hypertension is an excellent domain for the critiquing approach. (1) Hypertension is a chronic problem commonly encountered. (2) There is a large array of different drugs and different treatment modalities available (diuretics, beta blockers, central alpha agonists, vasodilators, angiotensin converting enzyme inhibitors, and calcium channel blockers), each with potential risks and benefits in particular patients. In addition, new agents are often added in these different categories. As a result, it is hard for the practicing physician to keep up to date and know how best to tailor his approach to a particular patient's care. (3) Relevant issues are periodically raised in the literature that can be brought to the physician's attention (e.g., the best dose for initiating a thiazide diuretic, as illustrated in example 1).

Using a system like HT-ATTENDING, a practitioner can get *specific feedback* as to how his style of practice fits in with current thinking in the field and with the changing spectrum of treatment alternatives that are available.

4.2. HT-ATTENDING System Design

To critique a physician's plan in a focused analysis, HT-ATTENDING's knowledge of hypertension management must be organized in a flexible fashion. The system must be able to individualize its comments not only to a particular patient, but to a particular approach to that patient's therapy.

This flexibility is built into the system design at three levels:

1. The individual comments from which the system's critique is built are stored in a format that allows their exact phrasing to vary depending on the overall context in which they are used.
2. The system structures its knowledge of treatment modalities hierarchically with individual comments associated at any level of the hierarchy.
3. When constructing its prose analysis, HT-ATTENDING uses "expressive frames" to help organize its comments and help make the train of thought flow smoothly from one comment to another.

The remainder of this section discusses these design features in turn. The initial version of HT-ATTENDING, described here, was not built using

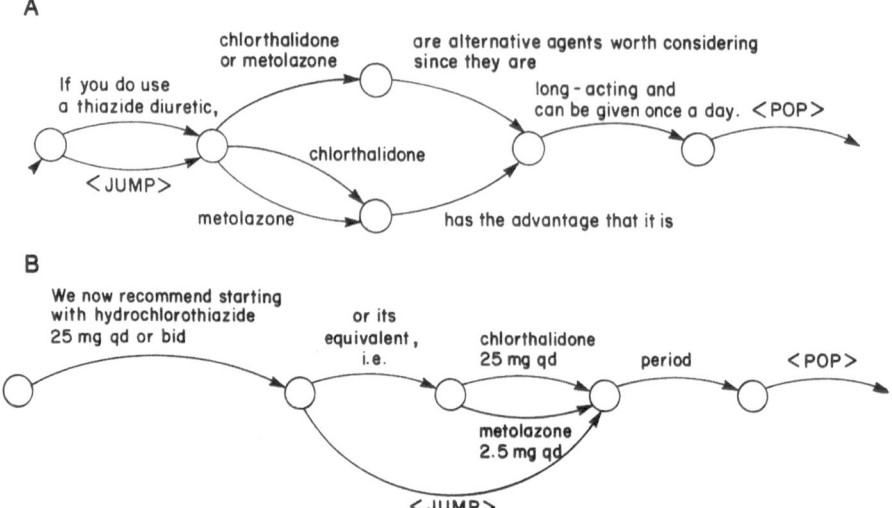

Figure 4.1. Two example comments expressed using PROSENET networks.

ESSENTIAL-ATTENDING (see Chapter 7). It was, however, later rewritten in the ESSENTIAL-ATTENDING format. This revision did not involve a great deal of change, since ESSENTIAL-ATTENDING's design evolved fairly directly from HT-ATTENDING's.

4.2.1. The Individual Comments

Depending on a patient's concurrent medical conditions and on the exact treatment proposed, a particular set of comments is incorporated into HT-ATTENDING's analysis. The exact phrasing of each of these comments may vary depending on the overall context in which it is used.

HT-ATTENDING achieves this flexibility using PROSENET, an approach developed to facilitate the generation of English prose analysis. Figure 4.1 shows how two comments are represented using PROSENET networks. As described in Chapter 7, PROSENET allows a great deal of flexibility in phrasing the prose critique so that it will sound natural to the reader.

4.2.2. A Hierarchy of Treatment Modalities

The second level of HT-ATTENDING's flexibility is its hierarchical structuring of the treatment modalities, as illustrated in Figure 4.2. In the case of diuretics, for instance, this hierarchy has four levels: diuretic, loop vs. thiazide, specific agent, and dose. Comments may be associated with any level of this hierarchy, discussing specific risks or benefits of using that treatment in a particular patient, or making remarks of a more general nature.

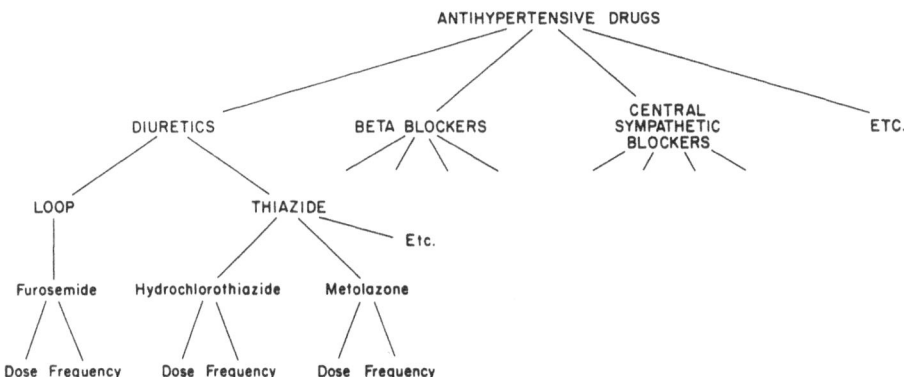

Figure 4.2. HT-ATTENDING's hierarchical structuring of treatment modalities.

For example, paragraph 2 of example 1 contains a comment discussing the possibility of a gouty attack when a diuretic is started in the presence of gout. This comment is associated at the diuretic level of the hierarchy since it applies to any diuretic. It is therefore output whenever any diuretic is started in a patient with gout. Had the comment applied only to thiazide diuretics, or only to hydrochlorothiazide, or only to a particular dose, then the comment would have been associated at the appropriate *lower level* in the hierarchy.

4.2.3. "Expressive Frames" Help Organize the Analysis

The previous subsections have described how HT-ATTENDING stores its individual comments in a way that leaves the exact phrasing flexible, and how it allows the comments to be hierarchically adapted to the physician's plan. The final requirement is that the comments be merged into a prose critique in which the thread of thought flows smoothly.

This overall organization of the critique is accomplished using "expressive frames." Figure 4.3 outlines the expressive frame for beta blockers. This frame contains all possible comments that might be made when use of a beta blocker is proposed. Associated with each comment is a variety of information, including a condition. If this condition is met, then the comment will be included in the analysis. The condition may be simple (e.g., "if the patient has asthma" or "if the patient is hypothyroid") or it may be more complex.

If the physician has proposed using a beta blocker, the system first examines all these conditions to see which comments should be included in the critique. It then arranges the comments in order by looking at certain other information stored with each comment. Finally, it generates the prose critique. In so doing, one further feature of the frame may be used. If necessary, as described in Section 7.8, a paragraph level PROSENET network is used to insert appropriate connective words and phrases around the comments, thereby further helping the train of thought to flow smoothly from one comment to another.

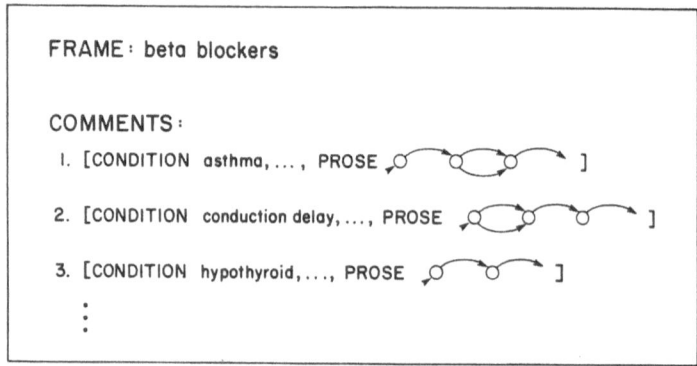

Figure 4.3. An example expressive frame containing information used when discussing beta blockers.

4.3. Discussion: Dealing with the Inherent Variation in Medical Practice

The large inherent variation in medical practice represents a major problem in the development of a medical computer-advisor. The critiquing approach allows HT-ATTENDING to deal with this variation as a central part of its design.

In dealing with practice variation, HT-ATTENDING must confront two quite different types of issues. First, there are a number of widely acknowledged management considerations involving the recognized risks and benefits of using certain agents with certain patients. Second, there is the less well-defined area of practice preferences.

In this regard, one might consider the practice of hypertension management to be a "structure" built up over years of clinical experience by many practitioners. This "structure" has two parts. First, there is a well-defined foundation of potential risks and benefits that forms the structure's underpinnings. Then, built on top of this well-defined foundation is a much more subjective, "softer" superstructure of practice preferences. To critique a physician's plan, HT-ATTENDING must be able to deal with both of these levels.

First let us consider the more well-defined management considerations.

1. Clear-cut management errors. There are a number of well-recognized management errors that may be proposed inadvertently, for instance, using a beta blocker in a patient with asthma, or giving a potassium-sparing diuretic to a patient with renal failure. In such a situation, as illustrated in example 2, HT-ATTENDING points out the problem and suggests reasonable alternatives.
2. Relative contraindications. Few risks involved in hypertension therapy, however, represent absolute contraindications. Most fall somewhere along a spectrum of relative contraindication. Examples are giving a beta blocker to

a patient with a history of congestive heart failure (CHF), giving a rauwolfia alkaloid to a patient with peptic ulcer disease, or giving a central alpha agonist to a patient with a history of depression. When such a treatment is proposed, HT-ATTENDING points out the problem but qualifies its comments appropriately, as shown for instance in paragraph 2 of example 3.

3. Beneficial aspects of a given approach. In addition to risks, there are also well-recognized benefits of using certain agents in certain patients. Examples are using a cardioselective beta blockers in a patient with peripheral vascular disease, or giving any beta blocker to a patient with coronary artery disease. Another very common benefit occurs because certain drugs are longer acting than others and can therefore be given less frequently. If the chosen approach has such a benefit, HT-ATTENDING can comment on that fact. If on the other hand, an alternative approach has benefits, the system can call these to the physician's attention.

The risks and benefits discussed above are well-recognized management considerations. Even here, however, there is ample room for practice variability. Different physicians may weigh the various risks differently. Also, the assessment of the magnitude of a given risk in a particular patient can be a very subjective judgement. In addition, some risks or benefits may be controversial, and others may not as yet be clearly established. In such a case, HT-ATTENDING can allude to the controversy or the uncertainty when mentioning the risk.

Even these well-recognized considerations, therefore, allow substantial latitude for practice variation. HT-ATTENDING's philosophy is to call these risks or benefits to the physician's attention and let him then do as he pleases.

Beyond these well-recognized, if occasionally controversial, considerations, HT-ATTENDING must also deal with the less well-defined issues discussed earlier. These "softer" issues include (1) "illogical" and unusual approaches, (2) practice preferences, and also (3) the significant variation that exists even in approaches published by authorities in the field.

1. Illogical and unusual combinations. Certain combinations of agents may make little sense. Examples are using two different thiazide diuretics, or two different beta blockers, at the same time. Here, the system can simply comment: "We do not see any advantage to using two different beta blockers at the same time," or "It would be a very unusual, resistant patient for whom it would make sense to use two such potent agents as guanethidine and minoxidil at the same time."

At other times, the utility of a proposed combination may not yet be established. In such a situation HT-ATTENDING comments, for instance: "You have suggested using captopril and prazosin at the same time. We are aware of no extensive reported experience that would allow us to evaluate this combination."

Of course, what is illogical to one practitioner may be routine practice to another. Indeed the line between what is illogical and what is merely a matter of preference may be hard to draw. Fortunately, this line need not be drawn in any absolute way, since the system's phrasing of its comments is shaded and softened if appropriate, and since the physician is in any case free to do as he pleases.

2. Practice preferences. As mentioned previously, HT-ATTENDING has built into it a "preferred" approach to hypertension management. This preferred approach reflects that of the clinic where the system is being developed. These preferences are outlined in Figure 4.4 and are discussed in more detail by Black (1983). HT-ATTENDING is programmed to express these preferences when appropriate.

Examples of these preferences are: (1) HT-ATTENDING prefers to initiate antihypertensive therapy with a diuretic. (This preference is shared with most authorities in the United States.) (2) HT-ATTENDING prefers to use a beta blocker as a step two agent, using central sympathetic blockers or captopril as a step two agent only for patients for whom beta blockers may be inappropriate.

As illustrated in several of the examples, the system does not hesitate to express these preferences when they are relevant. The rationale for incorporating such preferences into a computer advisor is discussed more fully later.

3. Accommodating other published approaches. In addition to expressing its own practice preferences, HT-ATTENDING should recognize when a physician has proposed an approach advocated by other authorities in the field and acknowledge this fact in its comments even if the approach does not coincide with its own preferences. For instance, if antihypertensive therapy is initiated with a beta blocker, the system comments as follows: "In Europe and in some centers in the United States, beta blockers are recommended as initial or step one agents, especially for patients under 50 years of age. Most authorities in the United States, however, recommend diuretics as step one therapy."

Figure 4.5 outlines a well-known, published approach to hypertension management (The 1980 Report of the Joint National Committee on Detection, Evaluation, and Treatment of High Blood Pressure). It would be hard to criti-

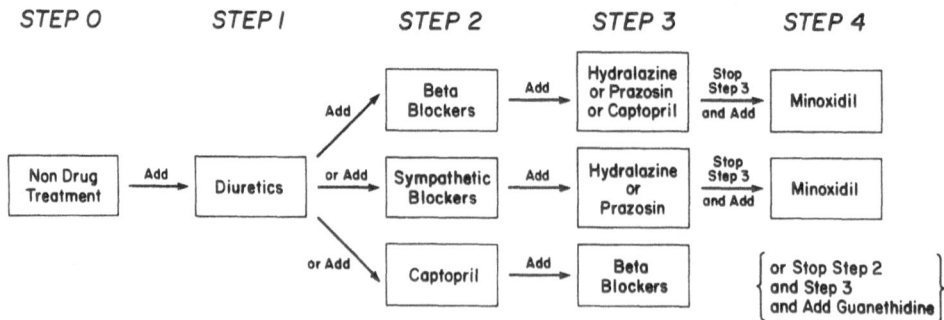

Figure 4.4. An outline of HT-ATTENDING's "preferred" approach to hypertension management (Black 1983).

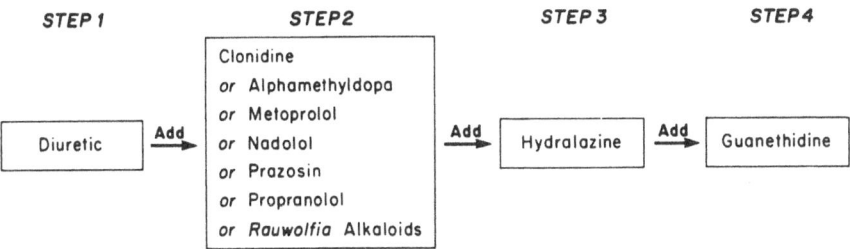

Figure 4.5. The stepped-care approach ot hypertension management outlined in the 1980 Report of the Joint National Committee on Detection, Evaluation, and Treatment of High Blood Pressure.

cize too severely a physician who was following this protocol, even if it was not HT-ATTENDING's preferred choice. Therefore, for instance, if a physician suggests starting proazosin as a step 2 drug, HT-ATTENDING comments: "Although we reserve prazosin to supplement the action of a beta blocker or a central adrenergic blocker, some authorities [1] do suggest its use as a step two drug in place of these agents." (Notice that in its comment the system references the published protocol. This citation is available online upon request.)

Similarly, if proazosin and hydralazine are suggested as a step 2 and step 3 drug respectively, HT-ATTENDING comments: "You have suggested using hydralazine and prazosin at the same time. Although a well known published protocol [1] technically allows this combination, we do not recommend it and prefer to use one or the other of these as a step three agent to supplement the action of a diuretic and a sympathetic blocker."

In summary, a number of distinctly different issues must be confronted by a computer-advisor in dealing with practice variability. This section has outlined how HT-ATTENDING deals with these issues. An unusual feature is that the computer is programmed with its own preferences and does not hesitate to express them when they are relevant to the case under discussion.

4.4. A Medical Computer-Advisor as a Form of "Publication" in Which an Expert Expresses His Views

At first glance it may seem somewhat strange to program a computer-advisor with its own set of practice preferences. On closer consideration, however, it is in many ways logical. Why should the computer be totally neutral in its advice? More to the point, why should the medical expert who designs the system be forced to be totally neutral?

In fact, the implementation of a medical computer-advisor may in time come to be accepted as a form of "publication." This publication might involve distribution of magnetic disks, or making the program available to practitioners via a nationwide computer network. In the future, instead of writing a journal article outlining his approach to a given problem, or in addition to doing so, a medical expert may write a computer-advisor to bring the same ideas to his audience in an interactive fashion, tailored to a particular patient's care.

Just as that expert would be expected to express his own practice preferences when writing the paper, it makes equal sense to allow him to incorporate these preferences into what is in effect the interactive version of that paper. In fact, HT-ATTENDING grew very directly from a paper on the evaluation and management of hypertension (Black 1983), written by our domain expert. With copyright permission, many sentences and phrases were included in the system. In addition, the overall philosophy in approach was incorporated as well. At the same time, of course, a substantial amount of additional information was incorporated to allow the system the flexibility and completeness described above.

As a result, HT-ATTENDING can be perceived as an "interactive paper" that tailors its content around a proposed approach to a patient's management. A physician could presumably get similar information by reading several articles on the subject. The advantage of using the computer is that the material is presented in a highly selective fashion and is focused around the problem at hand.

4.5. Finding the Most Efficient "Level of Detail"

Another interesting issue that a computer-advisor must face is the optimal level of detail at which to assist the physician. To be successful, the computer must minimize the time demands it places upon the clinician. A practitioner would likely not tolerate sitting at a terminal for 15 minutes or more answering a seemingly endless string of questions. One way to help streamline this interaction is by choosing the most "efficient" level of detail. Indeed, many issues need not be dissected in detail but can be alluded to and left to the physician's judgement.

An example is the choice of which diuretic to use when a patient has renal insufficiency. HT-ATTENDING's recommendation in this regard is that in a patient with "significant renal insufficiency (glomerular filtration rate less than 50 ml/minute)," a loop diuretic, such as furosemide, is most appropriate, otherwise a thiazide diuretic. One might design the system to ask questions evaluating kidney function (creatinine, creatinine clearance, etc.) before recommending a diuretic. This approach is definitely time consuming, possibly messy, at times subjective, but most important, totally unnecessary.

A much better approach is to defer the whole question to the physician. If a patient has renal insufficiency, HT-ATTENDING simply states it recommendation, explicitly defining its criteria (as illustrated in example 2), and leaves the physician to decide whether his patient has "significant renal insufficiency" or not. This approach avoids extra questions and still lets the system say succinctly everything it has to say.

Another example is starting a diuretic in a patient with gout. Here, the system need only indicate that this may precipitate a gouty attack, and that prophylactic treatment might be appropriate, as illustrated in example 1. HT-ATTENDING need not ask if, or how, the physician was planning to handle this problem or make any further comments on the subject.

4.6. Providing Further Information on Request

The heart of HT-ATTENDING is that part of the system which analyzes the proposed management plan. An attempt is made to keep this analysis succinct and to the point. On occasion, however, the physician may like additional details about issues mentioned in the critique. HT-ATTENDING allows the physician to ask for two types of further information.

1. Discussions of issues mentioned in the critique. HT-ATTENDING has a number of stored discussions, ranging in length from one paragraph to approximately one page, taken from a paper written by our domain expert (Black 1983). Example discussions are: "The Six Beta Blockers Presently Available in the United States," "Action and Choice of Diuretics," "Initiating Diuretic Therapy," "Captopril," "Guanethidine," "Minoxidil," etc. These serve to supplement the material presented in HT-ATTENDING's critique and are printed out upon request.
2. Online references. At the end of each critique, the physician can also ask for any reference citations. A short discussion of the paper may also be requested.

Although it is computationally simple to include these features, they may nevertheless significantly enhance the clinical value of the system.

4.7. HT-ATTENDING: Current Limitations

In implementing a medical computer-advisor, one must limit the scope of the problems the system will handle. For HT-ATTENDING, these limitations include:

1. The critique is focused on a useful but limited subset of hypertension management. HT-ATTENDING concentrates solely on antihypertensive agents used in outpatient treatment. It does not deal, for instance, with such in-hospital drugs as nitroprusside, nor is it designed to deal with hypertensive crises or emergencies. In fact, the system introduces itself with the following disclaimer:

> HT-ATTENDING is designed to assist in the pharmacologic management of essential hypertension on an outpatient basis. If your patient shows evidence of acute hypertensive end-organ disease, such as chest pain, hypertensive

encephalopathy, pulmonary edema, seizures, acute cerebrovascular compromise, or acute renal failure, then hospital admission is probably warranted for observation and possible parenteral therapy.

Thus HT-ATTENDING is primarily designed for use in a clinic or office setting. In this regard, the system's knowledge is fairly complete, including essentially all the antihypertensive agents and combination tablets listed in the 1983 Physician's Desk Reference.

2. The system's knowledge of conditions affecting hypertension mangement is bounded. HT-ATTENDING is familiar with 20 medical conditions that may affect hypertension management. Although these include the widely recognized factors, certain other conditions (such as pregnancy) are currently not included. As the system is refined, knowledge of a modest number of such additional factors may be added. Nevertheless, there will always be patients with uncommon medical problems with which HT-ATTENDING is unfamiliar, problems that may affect hypertension management. The system is not envisioned as an "automated reference" to help the physician learn the implications of rare diseases. Rather it is viewed as a tool to help him with the large majority of hypertensive patients whom he treats.

4.8. Preliminary System Use

Despite these limitations, a major current goal is to refine HT-ATTENDING to allow practical consultation. An important step in this direction will be to evaluate the system's performance in various ways.

The question of how best to evaluate a system that critiques medical management raises a number of interesting issues. With a system that performs *diagnosis*, one can input patient data and record how often the system's diagnosis is right or wrong. With a system that gives advice, it is not as clear how a quantitative evaluation is best performed. Indeed, the choices involved in hypertension management are quite subjective, which is why we advocate HT-ATTENDING's approach in the first place.

One might try to evaluate the system by recording how frequently it makes, misses, or catches clear-cut management errors, but this is a very small part of the system's performance. The real test is whether a physician finds the interaction productive, uses it again, and recommends it to his colleagues. These judgements are very subjective and heavily influenced by how comfortable a physician is with computer technology in the first place. Thus, just as the merit of a book is a subjective judgement of its readers, HT-ATTENDING's ultimate evaluation will occur as the system is used over time.

To gain some preliminary experience with the system in a clinical environment, we introduced HT-ATTENDING experimentally in our primary care clinic. Medical residents and nurse practitioners unfamiliar with the system were asked to submit interesting cases they had seen to test the system's performance.

Thirteen cases were contributed by different clinicians. For each case, the inter-action with the system typically took 3 to 4 minutes, divided roughly equally between typing information in and waiting while the critique was printed out. The system was familiar with the patients' major concurrent medical conditions and with the antihypertensive medications used.

The system suggested 37 treatment modifications for consideration, an average of 2.8 per patient. Seven of the alternatives involved agents whose main advan-tage was that they were longer acting than an agent proposed and might therefore be given less frequently. Examples of more substantive suggested alternatives included:

1. That hydrochlorothiazide be started at a dose of 25 mg rather than 50 mg
2. That a potassium-sparing agent be added when giving a diuretic to an insulin-dependent diabetic
3. That pindolol be considered as a beta blocker for a patient with congestive heart failure

In several patients, the clinicians volunteered that they would probably have changed the treatment based on the critique. The cases also indicated certain areas for further system refinement. Although this study is only preliminary, it does suggest that a system like HT-ATTENDING can offer a variety of construc-tive advice regarding patient care.

4.9. HT-ATTENDING: Summary

In summary, HT-ATTENDING extends our exploration of the critiquing approach into a very visible, important area of primary care. A current goal is to refine the system to the point where it can be disseminated as a practical consulta-tion tool. We feel that essential hypertension is a very promising domain for use-ful consultation. Two major steps taken toward this goal are:

1. HT-ATTENDING has been converted to the ESSENTIAL-ATTENDING for-mat, described in Chapter 7, which has given the system a cleaner, more organized internal structure. (As mentioned previously, this conversion did not involve extensive modifications, since ESSENTIAL-ATTENDING itself evolved from HT-ATTENDING's design.)
2. HT-ATTENDING is currently being updated to reflect current thinking in the field. In the 2 years since the initial system was developed, several new anti-hypertensive drugs have been approved for use. Also, the field has undergone a significant reformulation of certain aspects of the stepped-care approach, including a much broader advocacy of different drugs for initial, single-drug treatment. These major changes in a short period of time highlight the need (discussed in Chapter 8) for tools to help a system's designer keep a system's knowledge base up to date.

This chapter has described HT-ATTENDING's initial implementation and has outlined several design issues explored. It is worth reemphasizing, however, that a central conceptual contribution of HT-ATTENDING is the model it provides of an expert critiquing system as an *interactive paper*.

In fact, if a computer-advisor is presented to the physician as a computer-based "journal article" that tailors its content to his practice, this familiar model may help to demystify the computer's role. It may come to be seen as a natural addition to medical practice, paralleling the advances computers are making in many areas of everyday life.

Chapter 5

VQ-ATTENDING: Goal-Directed Critiquing of Ventilator Management*

The VQ-ATTENDING system extends the exploration of the critiquing approach into a third area of medical management. This domain has a considerably different character from previous domains. These differences have uncovered a new set of design issues.

VQ-ATTENDING is designed to critique aspects of a physician's ventilator management of a patient receiving mechanical respiratory support. In its current developmental implementation, VQ-ATTENDING explores a particular expert system-design feature: the ability to assess appropriate *treatment goals* and to use those goals to guide the system's critiquing analysis.

An implicit or explicit recognition of its goals must lie at the heart of the critique of any plan. In most existing expert systems, knowledge of treatment goals is buried *implicitly* in the system's logic. VQ-ATTENDING's implementation explores how underlying treatment goals may be made *explicit*. The system itself (1) infers a set of treatment goals that it considers appropriate for the patient described, (2) uses the goals internally to direct its critiquing analysis, and (3) discusses the goals in its prose critique of the physician's plan.

Thus VQ-ATTENDING explicitly separates *strategic* knowledge about treatment goals from *tactical* knowledge about management choices for achieving those goals. Ventilator management is a good domain for exploring such a "goal-directed" design, for reasons discussed in Section 5.6. It is anticipated that in appropriate domains, a goal-directed design may significantly enhance both an expert system's internal structure, and its ability to communicate clearly with its users.

---

* This chapter is adapted from P.L. Miller: Goal-directed critiquing by computer: Ventilator management. Computers and Biomedical Res. 18:422–438, 1985. Copyright 1985 by Academic Press, Inc. Reprinted by permission of the publisher.

5.1. Ventilator Management

The management of a patient receiving mechanical respiratory support is a broad domain. VQ-ATTENDING addresses a limited subproblem: *the feedback loop between arterial blood gas (ABG) data and ventilator settings*. The development of ABG analysis has made ventilator management a science, since it allows ventilator settings to be adjusted quantitatively to a patient's respiratory status (Shapiro et al. 1977).

Nevertheless, the process of choosing appropriate ventilator settings is often one of iterative trial and test. There are no fixed rules to follow, although there are general guiding principles. Individual patients respond differently, and physicians have different styles of practice. As a result, it makes sense (1) to let the computer help the physician optimize his own ventilator settings, and (2) to let the computer structure its advice from an appreciation of the underlying principles (treatment goals).

To use VQ-ATTENDING, the physician inputs the following information:

1. A small amount of medical information describing the patient, including certain such relevant medical conditions as increased intracranial pressure and low cardiac output
2. A current set of ABG results that are believed to be accurate, indicating pH, pO_2 (partial pressure of oxygen), and pCO_2 (partial pressure of carbon dioxide)
3. The current ventilator settings, as discussed below
4. A proposed set of new ventilator settings (which may of course be unchanged)

VQ-ATTENDING then critiques the appropriateness of the proposed settings, based on its assessment of appropriate treatment goals for the patient described, as illustrated in the examples of Section 5.2. The particular ventilator settings that the system critiques are:

1. FiO_2: fractional inspired oxygen, which can vary from 0.2 (room air) to 1.0 (100% oxygen)
2. PEEP: positive end-expiratory pressure, which may be applied to the breathing system to help improve a patient's oxygenation
3. RESPIRATORY RATE (RR): the minimum number of mechanically assisted breaths that the patient is to receive each minute
4. TIDAL VOLUME (TV): the volume of each mechanically assisted breath
5. MODE: the ventilator mode, either A/C (assist/control) mode or IMV (intermittent mandatory ventilation) mode
6. DEADSPACE: extra tubing (dead space) that may be inserted between the ventilator and the patient to help raise pCO_2.

These six variables represent the main parameters used with modern mechanical ventilators to manage a patient's respiratory status. There are certainly many other aspects in the care of a patient receiving mechanical respiratory support. VQ-ATTENDING focuses on a limited but central subproblem. This particular subproblem happens to lend itself well to exploring the system's goal-directed design. In fact, intensivists themselves have emphasized the goal-directed nature of respiratory care (e.g., Civetta 1983).

It is worth emphasizing that ventilator management has a considerably different character from previous domains in which the critiquing approach has been implemented. Previous critiquing domains have involved *discrete alternatives*: a decision might involve choosing either drug A, B, or C. In contrast, in ventilator management, important choices must be specified from a *continuum of possible values*. For instance, FiO$_2$ can vary from 0.2 to 1.0. PEEP can range from 0 cm H$_2$O to a practical maximum of 30 to 40. Similarly, respiratory rate and tidal volume are also chosen from a continuum of possible values. As discussed in Section 5.6, this continuum of choice may be why VQ-ATTENDING's goal-directed design is particularly appropriate.

5.1.1. VQ-ATTENDING vs. VM

The VM system (Fagan et al. 1979) is a developmental expert system also implemented to assist in ventilator management. VM and VQ-ATTENDING are quite different in the problems they address and from a clinical standpoint complement each other quite nicely. VM monitors real-time cardiovascular and ventilatory parameters and makes real-time treatment recommendations regarding ventilation. In contrast, VQ-ATTENDING is designed for intermittent consultation by a physician as he makes rounds. Another difference is that VM does not share VQ-ATTENDING's goal-directed design.

5.2. Four Examples

This section gives four examples that illustrate how VQ-ATTENDING critiques a proposed set of ventilator settings. In these examples, the arterial partial pressures of oxygen (pO$_2$) and of carbon dioxide (pCO$_2$) are given in units of *mm Hg*, PEEP in units of *cm H2O*, and tidal volume (TV) in *cc*. Dead space refers to *cc* of *extra* dead space added to the ventilator circuit.

Example 1. In this example, the following information has been input to the system by a physician:

PATIENT DESCRIPTION

age = 56
sex = M
weight (kg) = 60
arterial blood gas: pH = 7.27 pO$_2$ = 100 pCO$_2$ = 46
measured minute ventilation (L/min) = 7.2
measured respiratory rate = 9

VENTILATOR SETTINGS

	FIO2	TV	RR	PEEP	MODE	DEADSPACE
current:	0.6	800	9	12	AC	0
proposed:	0.4	800	10	12	AC	0

This information describes a patient whose arterial oxygenation is good ($pO_2 = 100$), and as a result, the physician is planning to reduce FiO_2 from 0.6 to 0.4 while leaving PEEP at 12. From the standpoint of ventilation the patient's pCO_2 is slightly high (40 is normal), and since the patient has a slight respiratory and metabolic acidosis (determined from the pH and pCO_2) it is probably desirable to have the pCO_2 lower than 40 to create a compensatory respiratory alkalosis to help normalize the pH.

The current implementation of VQ-ATTENDING produces the following critique of this plan. Notice that the critique is divided into two sections, discussing (1) oxygenation and (2) ventilation in turn. In each section, VQ-ATTENDING first outlines the treatment goals it considers to be relevant, and then critiques the suggested plan from the standpoint of those goals.

VQ-ATTENDING'S CRITIQUE

In regard to oxygenation, there are several goals for this patient's management. One goal is to maintain an adequate paO_2. A second goal is to reduce the risks associated with high PEEP. A third goal is to reduce the level of oxygenation support.

You have suggested a moderate decrease in FiO_2, to 0.4. Once an adequate oxygenation can be obtained with an FiO_2 of 0.6, we normally concentrate on lowering PEEP if it is high. We therefore would suggest an FiO_2 of 0.6 for the time being. You have proposed leaving PEEP at 12. A decrease in PEEP is probably possible.

Looking next at ventilation, the primary ventilatory goal for this patient is to achieve a moderate hypocarbia to compensate for the patient's metabolic acidosis.

You have proposed an increase in ventilation. If the patient's ventilation is completely controlled, the proposed increase in minute ventilation of 10%, from 7.2 to 8.0 L/m, should decrease the pCO_2 proportionally, to approximately 41. From the standpoint of the goal to achieve a moderate hypocarbia to compensate for the patient's metabolic acidosis, a somewhat larger decrease in pCO_2 may be desirable.

Example 2. This example involves a patient with several problems. His oxygenation (pO_2) is dangerously low. Also, there are conflicting goals in regard to ventilation. Section 5.6 discusses the problem of conflicting goals and how they may be dealt with in a critiquing system.

PATIENT DESCRIPTION

age = 67
sex = M
weight (kg) = 54
underlying diseases: increased intracranial pressure
arterial blood gas: pH = 7.60 pO_2 = 40 pCO_2 = 44
measured minute ventilation (L/min) = 6.4
measured respiratory rate = 8

VENTILATOR SETTINGS

	FIO2	TV	RR	PEEP	MODE	DEADSPACE
current:	0.6	800	8	10	AC	0
proposed:	0.6	800	11	12	AC	0

VQ-ATTENDING'S CRITIQUE

In regard to oxygenation, the primary goal in this patient's management is to achieve an adequate paO$_2$. Secondary goals are to avoid the risk of oxygen toxicity and to avoid the risks associated with high PEEP. The urgency of the primary goal, however, may override the secondary considerations.

You have suggested leaving the FiO$_2$ at 0.6. In view of the urgency of the patient's hypoxemia, it may be worthwhile to use an FiO$_2$ of 1.0, at least until the patient's oxygenation status has stabilized. Maintaining an FiO$_2$ greater than 0.6 over extended periods, however, may result in oxygen toxicity. You have proposed a small increase in PEEP. In a patient with increased intracranial pressure, increased PEEP may exacerbate this problem. In view of this patient's poor oxygenation, an increase in PEEP may ultimately prove necessary, but in this patient, you might first assess the effect of an FiO$_2$ of 1.0.

In regard to ventilation, there are conflicting goals for this patient's management. One would like to achieve a moderate hypocarbia to reduce cerebral blood flow. Also, a recommended goal is to achieve a moderate hypercarbia to compensate for the patient's metabolic alkalosis. The relative urgency of these problems and the optimal resolution of this conflict is outside the scope of this program and depends on the patient's clinical presentation, which should be carefully evaluated.

You have suggested an increase in ventilation. If the patient's ventilation is completely controlled, the proposed increase in minute ventilation of 37%, from 6.4 to 8.8 L/m, should decrease the pCO$_2$ proportionally, to approximately 32. From the standpoint of the goal to achieve a moderate hypocarbia to reduce cerebral blood flow, the decrease in pCO$_2$ which you suggest is appropriate.

On the other hand, from the standpoint of the goal to achieve a moderate hypercarbia to compensate for the patient's metabolic alkalosis, instead of the decrease in pCO$_2$ you suggest, a moderate increase may be appropriate.

Example 3. The following example involves a patient with quite good oxygenation who exhibits "primary hyperventilation," since he himself is driving the ventilator faster than its set rate.

PATIENT DESCRIPTION

age = 55
sex = M
weight (kg) = 74
arterial blood gas: pH = 7.58 pO$_2$ = 115 pCO$_2$ = 27
measured minute ventilation (L/min) = 9.5
measured respiratory rate = 12

VENTILATOR SETTINGS

	FIO2	TV	RR	PEEP	MODE	DEADSPACE
current:	0.5	800	8	5	AC	0
proposed:	0.5	800	6	5	AC	50

VQ-ATTENDING'S CRITIQUE

In regard to oxygenation, there are several goals for this patient's management. One goal is to maintain an adequate paO_2. A second goal is to reduce the level of oxygenation support. A third goal is to maintain PEEP at maintenance levels (0 to 5).

You have suggested leaving FiO_2 at 0.5. Once adequate oxygenation is obtained and FiO_2 and PEEP are not excessive, there is no pressing need for changes in oxygenation support until oxygenation significantly improves. Your decision to make no changes in FiO_2 is therefore reasonable. You have proposed leaving PEEP at 5. We agree that small amounts of PEEP may be appropriate to help prevent microatelectasis.

Turning now to ventilation, the primary ventilatory goal for this patient is to counteract, if possible, the patient's primary hyperventilation.

You have suggested a decrease in ventilation. If the patient's ventilation were completely controlled, the proposed decrease in minute ventilation should increase the pCO_2 proportionally, to approximately 36. The actual minute ventilation, however, is considerably higher than the ventilator settings. Since the patient's own ventilation seems to be driving the ventilator, it may be difficult to increase pCO_2. Although increasing dead space may increase pCO_2, it may just force the patient to breathe more and become exhausted. Most practitioners would not recommend the use of added dead space. Switching the mode of ventilation to IMV may be helpful. Similarly, decreasing tidal volume may help. If appropriate, it may be necessary to take other steps, such as sedating the patient.

Example 4. This final example shows a patient whose respiratory status appears to be quite good. This may be a patient whose pulmonary function is recovering, or who is being ventilated for some other reason, e.g., coma, flail chest, etc. Therefore, it may be appropriate to start weaning the patient from the ventilatory support.

PATIENT DESCRIPTION

age = 35
sex = F
weight (kg) = 58
arterial blood gas: pH = 7.52 pO_2 = 100 pCO_2 = 30
measured minute ventilation (L/min) = 9.5
measured respiratory rate = 10

VENTILATOR SETTINGS

	FIO2	TV	RR	PEEP	MODE	DEADSPACE
current:	0.4	750	10	5	AC	0
proposed:	0.4	750	10	0	AC	0

VQ-ATTENDING'S CRITIQUE

In regard to oxygenation, there are several goals for this patient's management. One goal is to maintain an adequate paO_2. A second goal is to maintain FiO_2 at maintenance levels (0.2 to 0.4). A third goal is to maintain PEEP at maintenance levels (0 to 5).

You have suggested leaving the FiO_2 at 0.4. Once good oxygenation is attained and both FiO_2 and PEEP are at ambient levels, any further changes are largely optional. Your decision to make no changes in FiO_2 is therefore reasonable. You have proposed a moderate decrease in PEEP. Some practitioners recommend that small amounts of PEEP (i.e., 3 to 5) may help in preventing microatelectasis.

Looking next at ventilation, in the absence of problems such as significant metabolic acidosis or alkalosis or increased intracranial pressure, the primary goal for this patient is to reduce the level of ventilatory support.

You have suggested leaving the patient's minute ventilation unchanged. In the absence of other medical problems which make ventilatory weaning inappropriate, a reduction in ventilatory support may be possible. If you want to start reducing the level of ventilatory support, changing to IMV mode is a reasonable step.

These critiques illustrate at the surface level that VQ-ATTENDING uses two levels of interacting knowledge: (1) knowledge about *treatment goals*, and (2) knowledge about *management alternatives*. As a result, VQ-ATTENDING has a goal-directed character, both in its critiquing advice, as shown here, and in its internal analysis of the physician's plan as discussed in Sections 5.4 and 5.5.

5.3. VQ-ATTENDING: System Overview

Figure 5.1 shows a schematic overview of VQ-ATTENDING's internal system design. VQ-ATTENDING is implemented in the ESSENTIAL-ATTENDING format, described in Chapter 7. The system's analysis consists of four stages.

1. Information about a patient is gathered from the physician, together with the current and proposed ventilator settings.

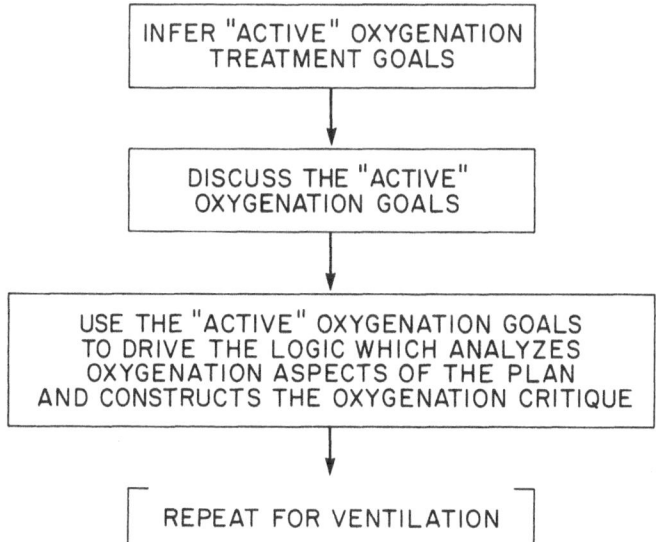

Figure 5.1. VQ-ATTENDING: Overview of the System Design.

2. From this information, VQ-ATTENDING then infers two sets of treatment goals (oxygenation goals and ventilation goals) that it considers relevant to the patient's management. These goals are inferred using production rules (Buchanan and Shortliffe 1984). Different sets of goals are inferred for different patients, depending on the severity of disease and the current level of ventilatory support.
3. Next a prose discussion of the inferred *oxygenation* goals is produced. This discussion is followed by a prose critique of the oxygenation component of the plan from the perspective of these goals. These prose discussions are produced using PROSENET (see Chapter 7), an approach developed to facilitate the machine generation of polished prose.
4. Finally, a similar prose discussion of *ventilation* goals is produced, followed by a prose critique of the ventilation component of the plan.

The following two sections describe the internal design of VQ-ATTENDING in detail. To implement a goal-directed system, one must (1) define the various goals of therapy present at different times in a patient's management, (2) specify the conditions that cause each goal to be active, and (3) indicate how the different goals affect management decisions. In VQ-ATTENDING, this process is done twice: first for oxygenation, then for ventilation.

5.4. A Goal-Directed System Design: Oxygenation

This section describes the internal design of that part of VQ-ATTENDING that critiques oxygenation. Central to this design are the system's oxygenation goals.

5.4.1. Oxygenation Goals

VQ-ATTENDING currently includes the following treatment goals in managing a patient's oxygenation:

Goal 1. To achieve an adequate arterial oxygenation (PO_2)
Goal 2. To maintain an adequate PO_2
Goal 3. To avoid the risk of potential oxygen toxicity
Goal 4. To reduce the risk of potential oxygen toxicity
Goal 5. To avoid the risks associated with high PEEP
Goal 6. To reduce the risks associated with high PEEP
Goal 7. To reduce the level of oxygenation support
Goal 8. To maintain FiO_2 at "maintenance" levels
Goal 9. To maintain PEEP at "maintenance" levels

In a given patient, a subset of these nine goals is active at one time, depending on the state of pulmonary function and the current level of respiratory support. For instance:

1. When a patient is initially placed on the ventilator with poor oxygenation, goals 1, 3, and 5 might be active.
2. Once stable oxygenation is achieved, goals 2, 4, and 6 might be active.
3. Once the patient's pulmonary disease improves, goals 2, 6, and 7 might be active.

These goals can be thought of as "active treatment considerations." It is important to note that not all active goals may be achievable. Indeed, active goals may conflict with one another. (The nature of these goals is discussed more fully in Section 5.6.)

5.4.2. Production Rules Infer "Active" Goals

VQ-ATTENDING's first step in analyzing the proposed management of the patient's oxygenation is to identify those oxygenation goals which apply to the patient described. These "active" goals are inferred by production rules, as illustrated below:

```
(RULE ACH_PO2_1
  (IF (SAME PO2 POOR))
  (THEN (ACHIEVE_PO2_GOAL ACTIVE)
        (ACHIEVE_PO2_GOAL PRIORITY HIGH)))

(RULE ACH_PO2_2
  (IF (SAME PO2 MARGINAL))
  (THEN (ACHIEVE_PO2_GOAL ACTIVE)
        (ACHIEVE_PO2_GOAL PRIORITY MOD)))

(RULE MAINT_PO2
  (IF (OR (SAME PO2 ADEQUATE)
          (SAME PO2 GOOD)
          (SAME PO2 EXCELLENT)))
  (THEN (MAINT_PO2_GOAL ACTIVE)))

(RULE REDUCE_O2_TOX
  (IF (AND (NOT (SAME PO2 POOR))
           (GREATERP FIO2CURR .6)))
  (THEN (REDUCE_O2_TOX_GOAL ACTIVE)))

(RULE RED_OX_SUPPORT
  (IF (AND (OR (SAME PO2 GOOD)
               (SAME PO2 EXCELLENT))
           (OR (GREATERP FIO2CURR .4)
               (GREATERP PEEPCURR 5))))
  (THEN (REDUCE_O2_SUPPORT_GOAL ACTIVE)))
```

Each of these example production rules is an "IF . . . THEN . . ." construct. In the case of the rules shown above, each IF clause tests aspects of the patient description, including the adequacy of oxygenation and current ventilator settings, and each THEN clause asserts a treatment goal as being active.

The meaning of each rule is fairly evident from the rule itself. For instance, the English translation of the first rule is: "If the patient's arterial oxygenation is poor, then the goal 'to achieve an adequate arterial oxygenation' is active and its priority is high." Similarly, the fifth rule translates: "If the patient's arterial oxygenation is good or excellent, and the current FiO_2 is greater than 0.4 or the current PEEP is greater than 5, then the goal 'to reduce the level of oxygenation support' is active."

The assignment of the patient's arterial oxygenation to a rough qualitative range (POOR, MARGINAL, ADEQUATE, GOOD, EXCELLENT) is based on the arterial pO_2 and the level of PEEP. This assignment is somewhat arbitrary, however, especially at the borders of each range. A potential problem may occur if, for instance, the pO_2 deemed "acceptable" by the system is deemed "unacceptable" by the physician. This is an example of a general problem that any expert system faces. As discussed in Section 5.6.5, an interesting feature of an explicitly goal-directed design is that the system may deal with such a conflict by critiquing *at the level of goals*, as well as at the level of management choices.

5.4.3. Critiquing the Oxygenation Plan

Once a set of active oxygenation goals has been inferred using production rules, as described above, they are incorporated into a prose discussion which is output as part of the critique. Finally, the active goals are used to guide the creation of VQ-ATTENDING's critiquing analysis. Central to this process, are the two ESSENTIAL-ATTENDING "expressive frames," shown below, which indicate the various prose comments that may be included in the critique:

```
(DEFFRAME 'OXCRITIQUE '(

(IF T COMMENT $OXCRIT SEQUENCE 0)

(IF (SAME ACHIEVE_PO2_GOAL PRIORITY HIGH) COMMENT $OX_HIGH
        SEQUENCE 1)

(IF (SAME ACHIEVE_PO2_GOAL PRIORITY MOD) COMMENT $OX_MOD
        SEQUENCE 1)

(IF (OR (SAME AVOID_O2_TOX_GOAL ACTIVE)
        (SAME REDUCE_O2_TOX_GOAL ACTIVE)) COMMENT $OX_TOX
        SEQUENCE 5)

(IF (AND (OR (SAME MAINT_PO2_GOAL ACTIVE)
            (SAME REDUCE _O2_SUPPORT_GOAL ACTIVE))
        (SAME REDUCE_O2_TOX_GOAL ACTIVE))
    COMMENT $OXMT1 SEQUENCE 1)
```

```
(IF (AND (OR (SAME MAINT_PO2_GOAL ACTIVE)
             (SAME REDUCE_O2_SUPPORT_GOAL ACTIVE))
         (NOT (SAME REDUCE_O2_TOX_GOAL ACTIVE))
         (SAME REDUCE_EXC_PEEP_GOAL ACTIVE))
    COMMENT $OXMT2 SEQUENCE 1)

(IF (AND (OR (SAME MAINT_PO2_GOAL ACTIVE)
             (SAME REDUCE_O2_SUPPORT_GOAL ACTIVE))
         (NOT (SAME REDUCE_O2_TOX_GOAL ACTIVE))
         (NOT (SAME REDUCE_EXC_PEEP_GOAL ACTIVE)))
    COMMENT $OXMT3 SEQUENCE 1)

(IF (AND (SAME MAINT_FIO2_GOAL ACTIVE)(SAME MAINT_PEEP_GOAL
    ACTIVE))
    COMMENT $MAMB SEQUENCE 1)

) 'GENCOMMENTS)


(DEFFRAME 'PEEPCRITIQUE '(

(IF T COMMENT $PDESC SEQUENCE 0)

(IF (SAME ACHIEVE_PO2_GOAL PRIORITY HIGH) COMMENT $PHIGH
    SEQUENCE 1)

(IF SAME MAINT_PEEP_GOAL ACTIVE) COMMENT $PAMB SEQUENCE 1)

(IF (NOT (OR (SAME ACHIEVE_PO2_GOAL PRIORITY HIGH)
             (SAME MAINT_PEEP_GOAL ACTIVE)))
    COMMENT $PEEPCRIT SEQUENCE 1)

(IF T COMMENT $PEEPGENT SEQUENCE 5)

) 'GENCOMMENTS)
```

The OXCRITIQUE expressive frame generates the discussion of FiO$_2$, and the PEEPCRITIQUE frame generates the discussion of PEEP. Each frame consists of a list of "comment frames," each of which indicates a possible comment that can be included in the critique. For instance, the third comment frame in OXCRITIQUE is:

```
(IF (SAME ACHIEVE_PO2_GOAL PRIORITY MOD) COMMENT $OX_MOD
    SEQUENCE 1)
```

This comment-frame consists of three parts:

1. A condition. The condition in this comment frame is "if the goal 'to achieve an adequate arterial oxygenation' is active with moderate priority." If this condition is true, then this comment will be included in the critique.
2. A pointer to the prose comment itself. The prose comment itself (pointed to by $OX_MOD) is expressed as a PROSENET network as illustrated below.

If this comment is selected to be output, then this PROSENET network will
be activated.
3. A sequence indicator. This indicates the sequence in which the chosen comments are to be output.

Notice that almost all of the 13 comment frames in these two frames contain
conditions that test the presence or absence of one or more treatment goals. *As
a result, the set of comments included in the oxygenation critique is totally controlled by the set of active goals inferred by the production rules.* This is the heart
of VQ-ATTENDING's goal-directed design. Also, as illustrated below, the active
goals may also influence how individual comments are phrased.

As described above, each comment frame contains a pointer to a prose comment. The comment itself is stored in the PROSENET format (see Chapter 7).
The PROSENET comment pointed to by $OX_MOD is shown below.

```
($OX_MOD ((we agree that an increase in FiO2 is appropriate *period)
              $POPTT (AND (LESSP FIO2CURR .6)(GREATERP FIO2PROP
              FIO2CURR)))

         ((instead of decreasing FiO2 *comma we would suggest a)
              $OXM4O (AND (LESSP FIO2CURR .6)(LESSP FIO2PROP
              FIO2CURR)))

         ((instead of leaving the FiO2 unchanged *comma we would suggest a)
              $OXM4O (AND (LESSP FIO2CURR .6)(EQUAL FIO2CURR
              FIO2PROP)))

         ((we agree that leaving the FiO2 at "0.6" is appropriate
         *period)
              $POPTT (AND (EQUAL FIO2CURR .6)(EQUAL FIO2PROP .6)))

         ((although it is desirable to decrease FiO2 eventually *comma)
              $OXM1O (SAME REDUCE_O2_TOX_GOAL ACTIVE))

         (*jump $OXM1O T)

         )

($OXM1O ((we agree that leaving the FiO2 unchanged is appropriate)
              $OXM2O (EQUAL FIO2CURR FIO2PROP))

         ((we would probably leave the FiO2 unchanged for now *comma)
              $OXM2O T)

($OXM2O ((until adequate oxygenation is achieved *period) $POPTT T))

($OXM4O ((*function (DESCRECFIO2CH)) $OXM5O T))

($OXM5O ((instead *period) $POPTT T))  .
```

The PROSENET approach is described in Chapter 7. PROSENET allows the
individual comments to be stored in a very flexible form, which in turn allows the

prose expression of a comment to be flexibly adapted to the exact set of circumstances being described.

Notice that the fifth arc of $OX_MOD includes a test as to whether a particular goal is active. Only if the goal is active will the arc's prose fragment be included in the critique. In this way, the set of active goals can directly influence the *phrasing* of an individual comment.

Thus the set of active treatment goals (1) intimately controls the choice of comments to be included in the oxygenation critique and (2) may also influence the prose phrasing of individual comments as well.

5.5. A Goal-Directed System Design: Ventilation

The design of the ventilation component of VQ-ATTENDING is similar to that of the oxygenation component described above.

5.5.1. Ventilation Goals

VQ-ATTENDING currently includes eight ventilation goals:

Goal 1. To maintain a normal pCO_2 and normal work of breathing
Goal 2. To achieve a normal pCO_2 and normal work of breathing
Goal 3. To maintain a moderate hypocarbia
Goal 4. To achieve a moderate hypocarbia
Goal 5. To maintain a moderate hypercarbia
Goal 6. To achieve a moderate hypercarbia
Goal 7. To counteract, if possible, the patient's primary hyperventilation
Goal 8. To reduce the level of ventilatory support

An interesting feature of these goals is that they are mostly mutually exclusive. In contrast with the oxygenation goals, only one ventilation goal usually applies at any one time. (As illustrated in example 2, however, the possibility of conflicting goals does exist and is discussed further in Section 5.6.4.)

Here again, VQ-ATTENDING uses a set of production rules to infer active ventilation goals, for example:

```
(RULE MAINT_HYPO
        (IF (AND
                (OR (SAME HX IICP)
                    (SAME ABG ACIDOSIS))
                (SAME PCO2 HYPOCARB MODERATE)))
            (THEN (MAINT_HYPOCARB_GOAL ACTIVE)))

(RULE ACH_HYPO
        (IF (AND
                (OR (SAME HX IICP)
                    (SAME ABG ACIDOSIS))
```

```
                  (NOT (SAME PCO2 HYPOCARB MODERATE))))
                  (THEN (ACHIEVE_HYPOCARB_GOAL ACTIVE)))
```

The second of these rules translates: "If the patient has increased intracranial pressure or if the arterial blood gas shows a metabolic acidosis and the pCO_2 is not moderately hypocarbic, then an active goal is 'to achieve a moderate hypocarbia.' "

After VQ-ATTENDING infers a set of active ventilation goals, and generates a prose discussion of those goals, then its critique of ventilation is constructed, in much the same fashion as described above in Section 5.4. Here again, an expressive frame (VENTCRITIQUE) indicates the comments that may be output.

```
(DEFFRAME 'VENTCRITIQUE

(IF T COMMENT $VCRIT SEQUENCE 0)

(IF (NOT (EQUAL MVCURR MCPROP)) COMMENT $V_SEQUENCE 1)

(IF T COMMENT $VCRIT2 SEQUENCE 2)

(IF (AND DSPACECURR (GREATERP DSPACECURR 0)(NOT (SAME VENT
    DRIVEN)))
        COMMENT $DSPACE SEQUENCE 4)

(IF (AND (EQ MODECURR 'IMV)(LESSP MVPROP MVCURR)(LE RATEPROP
    8)  (NOT  (SAME  REDUCE_VENT_SUPPORT_GOAL  ACTIVE)))
        COMMENT $NOWEAN SEQUENCE 4)

(IF (SAME REDUCE_VENT_SUPPORT_GOAL ACTIVE) COMMENT $WEAN
    SEQUENCE 6)

(IF (SAME REDUCE_VENT_SUPPORT_GOAL ACTIVE) COMMENT $MODECH
    SEQUENCE 7)

) 'VENTCRIT)
```

Here the set of active goals plays a less central role in determining which comments are to be used in critiquing ventilation. This is because the goals are largely mutually exclusive and are therefore more straightforward to coordinate when generating the critique. Nevertheless the treatment goals are still an integral part of the critiquing process.

This concludes the discussion of VQ-ATTENDING's internal design. The discussion has focussed primarily on the system's use of treatment goals to enhance its critiquing structure. It is worth mentioning, however, that to assist in this process, the system incorporates a variety of conventional equations governing respiratory physiology to perform such calculations as:

1. Estimating the change in pCO_2 produced by a change in minute ventilation
2. Approximating the change in pO_2 produced by a change in FiO_2
3. Determining the respiratory and metabolic acid/base components of an arterial blood gas result

These equations represent a very small part of the system, comprising only a few lines of LISP in a total program approximately 35 pages in length.

5.6. Goal-Directed Critiquing in Perspective

The explicit internal recognition of treatment goals may be particularly useful for a critiquing system, in appropriate domains. A critiquing system must not only be able to propose a good plan but must react to any plan (good, marginal, or poor). An ability to assess treatment goals explicitly may help to focus the system's analysis appropriately.

A brief discussion of VQ-ATTENDING's development may be of interest in this regard. In a previous implementation, the system's knowledge of treatment goals and of management alternatives was intermixed, and the distinction between these two types of knowledge was not recognized. This created a number of problems in the iterative process of fine tuning the system. Fine tuning for one set of circumstances often resulted in the system's starting to make inappropriate recommendations in other circumstances.

In retrospect, one problem was that the system's knowledge of treatment goals was incomplete. In fact, the number of implicit goals included in the previous implementation of the system (to the extent that they could be identified) was smaller than the number of goals currently identified. It is useful, for instance, to define separate goals for achieving a state (e.g., adequate arterial oxygenation) and for maintaining that state. These are quite distinct issues.

The remainder of Section 5.6 discusses VQ-ATTENDING's use of treatment goals to help define more clearly what they are, and in what domains an explicit recognition of treatment goals may enhance an expert system's design.

5.6.1. What Makes a Good Domain for Goal-Directed Critiquing?

An interesting question is whether or not VQ-ATTENDING's goal-directed design is generally applicable to a range of potential domains. As discussed in Section 5.1, ventilator management differs from previous critiquing domains in significant ways. Whereas in previous domains, management alternatives involved discrete choices (e.g., drug A, B, or C), in ventilator management there is a continuum of choice.

With discrete alternatives, one can assign specific, discrete risks and benefits to each choice in a particular patient. The decision process then is largely reduced to weighing the various risks and benefits against one another, as is done by the ATTENDING system in anesthetic management. In so doing, *specific treatment goals may be dealt with only implicitly in pursuing a broader goal of "minimizing overall risk."*

With a continuum of choice, it is impossible to assign clearly differentiated risks and benefits to alternatives. The risks and benefits themselves vary incrementally along a continuum, e.g., the higher the FiO_2, the higher the risk of oxygen toxicity; the higher the PEEP, the higher the risk of barotrauma and other complications. One cannot reduce the problem to discrete alternatives, each with discrete risks and benefits. As a result, the problem becomes fuzzier. Probably for this reason, ventilator management lends itself well to a goal-directed approach. The goal-directed design helps bring order to an otherwise loosely structured domain.

Looking beyond the domain of ventilator management, however, a broader question is how generally useful a goal-directed design will prove to be, and in what domains it will be best applied. The remainder of this section attempts to answer this question.

In any domain of medical management, there are always underlying principles. In many domains, however, these underlying principles may be more relevant to pharmaceutical companies developing new drugs than to the clinical practitioner. The practitioner often operates with a set of discrete alternatives, structured into official, semiofficial, or personal protocols, which are then adapted to individual patients.

Looking at other critiquing domains, for instance:

1. In oncology (ONCOCIN) there are elaborate official protocols that are followed.
2. In the management of essential hypertension (HT-ATTENDING), there are semiofficial stepped-care guidelines.
3. In anesthesiology (ATTENDING), there is a well-defined sequence of premedication, induction, intubation, and maintenance, with established agents and techniques for each.

Thus the physician frequently practices within the limits of established guidelines involving discrete alternatives, and his management decisions involve choosing among these so as to minimize risk. Only if the patient is unusual in some respect may a physician fall back to more fundamental reasoning.

In summary, established patterns of care have evolved over years of clinical practice. When a patient fits this established pattern, falling back to underlying treatment goals may be unnecessary. There is no need to reinvent the wheel. When a patient does not fit the standard pattern, however, or when the domain itself (e.g., ventilator management) makes it difficult to establish discrete choices, then a more goal-directed approach may be most useful.

Another domain where the goal-directed design might be appropriate is the problem of titrating insulin to control a diabetic patient's blood sugar. Here, the doses of each type of insulin must be chosen from a continuum of possible values. Also, although there are no explicit rules to guide treatment, there are several clear treatment goals: to avoid hypoglycemia, to achieve/maintain adequate control of blood glucose, to reduce/maintain the morning fasting blood glucose to an appropriate level, to minimize the number of daily doses, etc. An explicit recognition of these goals might much enhance a system designed to critique insulin therapy.

5.6.2. Treatment Goals vs. Inferencing Goals

A potential source of confusion regarding VQ-ATTENDING's use of *treatment goals* is that a number of other expert systems, e.g., MYCIN (Shortliffe 1976), use an entirely different type of "goal" (*inferencing* goals) in their internal analysis. In particular, when such a system uses its production rules in a "backwards-chaining" fashion, the conclusions indicated by the THEN clause of a rule become *inferencing goals*. The system attempts to confirm these by investigating the various conditions in the IF clause of the rule, which themselves then become inferencing subgoals.

It is worth emphasizing that the *medical* treatment goals used by VQ-ATTENDING are totally different from *expert system* inferencing goals. Treatment goals are medical principles that guide therapy. In fact, throughout this chapter, the term "treatment goal" is usually used when discussing these goals to help avoid this potential confusion.

5.6.3. VQ-ATTENDING's Treatment Goals: Are They Clinical or Computer Entities?

It is difficult to define the exact nature of VQ-ATTENDING's treatment goals in a rigorous, formal way. The goals might best be thought of as "treatment considerations" or "underlying treatment principles." They are certainly clinical entities. They are relevant and natural from the perspective of the patient's clinical care.

On the other hand, the primary function of these goals is not clinical per se, but rather to enhance the internal structure of an expert system's critiquing logic. The particular set of goals used is chosen to maximize this expert system design function. As a result, the set of goals used may not necessarily be seen as comprehensive by a clinical expert in the domain.

5.6.4. Dealing with "Conflicting Goals"

Any system that deals with treatment goals must be prepared to deal with conflicts between the goals. In VQ-ATTENDING there are two general types of such conflicts:

1. Conflicts where it is clear which goal has priority. It may happen that several goals are active but not all can be achieved. In many circumstances, one can tell a priori which is most important. An example of this type of conflict is seen in the oxygenation critique of example 2. Here, the need to achieve adequate oxygenation overrides two other active goals. In fact, this priority is explicitly mentioned in the paragraph discussing oxygenation goals.

   The priority of this goal is built into the logic that generates the critique. This is possible because it was known when building the system that that goal would always have priority. (If conflicting goals could have *different* priorities depending on the patient's condition, of course, then the critiquing logic would have to be more flexible.)

2. Conflicts where it is unclear which goal has priority. Another possibility is that it might be unclear which of two conflicting goals should take precedence. An example is seen in the ventilation critique of example 2. Here, there are two conflicting goals as to the appropriate level of pCO_2, since the patient has (1) increased intracranial pressure and (2) a significant metabolic alkalosis.

As discussed in the critique, this conflict can be resolved only on the basis of complex factors outside the scope of the system's expertise. As a result, the system alerts the physician to the conflict, critiques his plan independently from the standpoint of each goal, and then leaves the final resolution of the problem to the physician.

In summary, the problem of conflicting goals is a natural outgrowth of a goal-directed design. The problem can be dealt with in different ways depending on the character of the domain and on the goals themselves.

5.6.5. Critiquing at the Level of Goals

In its current implementation, VQ-ATTENDING infers a set of treatment goals that *it* considers relevant to the patient's care and then critiques the physician's proposed management from the standpoint of those goals. An interesting refinement would be to also allow the system to *critique the physician's choice of treatment goals*. This possible extension of a goal-directed critiquing system is discussed in more detail in Section 9.4.3.

The incorporation of a capability to critique treatment goals would be a major research project in its own right. It is worth emphasizing, however, that the issues involved are equally relevant whether or not a system has a goal-directed design. In a system where the distinction between goals and management alternatives is not made, however, the distinction between critiquing management choices and critiquing goals would be totally lost. By separating *stategic* knowledge about treatment goals from *tactical* knowledge about management choices, it becomes clear that two types of critiquing with very different character, can be done.

5.7. VQ-ATTENDING: Limitations

The current implementation of VQ-ATTENDING is limited in a number of ways. First of all, as mentioned in Section 5.1, the system deals with a central but limited subproblem in the broader field of managing a patient receiving respiratory support.

The largest present limitation, however, is that VQ-ATTENDING is a developmental system. Although it performs reasonably for a range of representative test cases, it has not been subjected to rigorous evaluation. The goal of the current implementation is experimental: to explore the particular design issues

involved in incorporating a goal-directed control structure into an expert critiquing system.

5.8. Summary: The Separation of Strategic from Tactical Knowledge in an Expert System

In appropriate domains, the explicit definition of treatment goals and the separation of strategic knowledge about these goals from the tactical knowledge of
management alternatives may offer several advantages:

1. The explicit identification of treatment goals may allow a more comprehensive set of goals to be defined.
2. The separation of knowledge that determines which treatment goals are active
   from knowledge about management itself may enhance the system's logical
   structure.
3. This, in turn, may help in fine tuning the system and may facilitate discussion
   of the system's knowledge with domain experts.
4. The goal orientation may have particular value for a critiquing system that
   must respond appropriately to *any* management plan and that must have its
   knowledge available to it in a flexible form.

   In many areas of life, it is useful to define one's goals explicitly when embarking on a project. It is therefore not surprising that this paradigm may also be
reflected in an expert system's internal design. The ability to structure a critiquing analysis around an explicit definition and assessment of treatment goals
has an intuitive appeal. The current development of VQ-ATTENDING is a
step in exploring how this paradigm might best be incorporated into an expert
computer-advisor.

Chapter 6

PHEO-ATTENDING:
Pheochromocytoma Workup and Conflicting Expertise*

The PHEO-ATTENDING system extends the exploration of the critiquing approach beyond medical management into an area of medical workup. PHEO-ATTENDING is designed to critique a physician's workup of a patient with a suspected pheochromocytoma. To use PHEO-ATTENDING, a physician first inputs a modest amount of information describing the patient and indicates any tests or procedures that he has ordered, or that he plans to order. He also specifies the results of any tests and procedures that have already been performed.

PHEO-ATTENDING then critiques the physician's workup, discussing its appropriateness for the patient described, and mentioning any other approaches that might be preferred. The system may be consulted at any point during the workup process, which may involve a sequence of tests. In its critique, PHEO-ATTENDING gives the physician feedback as to how his approach fits in with the current spectrum of tests and procedures available.

PHEO-ATTENDING has been implemented as a developmental prototype system. The goal of the present research is to explore the general design issues involved in building a computer system that critiques workup. A major focus is the problem of how best to incorporate *conflicting expertise*, which occurs when domain experts themselves advocate different approaches to a problem. It is anticipated that the system design developed for pheochromocytoma can be extended to other areas of patient workup.

PHEO-ATTENDING was developed in collaboration with Dr. Henry R. Black, a hypertension specialist. The system itself was implemented by Steven J. Blumenfrucht, a Yale medical student, as his M.D. thesis project.

*This chapter is adapted from P.L. Miller, S.J. Blumenfrucht, and H.R. Black: An expert system which critiques patient workup: Modeling conflicting expertise. Computers and Biomedical Res. 17:554–569, 1984. Copyright 1984 by Academic Press, Inc. Reprinted by permission of the publisher.

6.1. Critiquing Patient Workup

Patient workup is an area of medicine that lies between diagnosis and treatment. Workup starts once a physician has developed a differential diagnosis based on information gathered from history, physical examination, and initial laboratory tests. The differential diagnosis lists possible problems a patient may or may not have. A physician must then order further tests and procedures to rule in or rule out these problems, or to assess their character and severity.

One characteristic that workup shares with many other areas of medicine is that it is not static. The optimal strategy for working up a problem evolves as new tests and procedures are developed, and as new studies are performed evaluating their relative efficacy. As each new test is introduced, there may be a period of controversy while its proper role becomes established. It may become widely used, either as a general tool or in specific circumstances, or it may pass into disuse. In the face of this ongoing evolution, the critiquing approach is particularly appropriate.

Despite the importance of patient workup, however, little expert system research has specifically addressed the problems involved. To the extent that previous expert systems have dealt with workup, it has usually been in diagnostic systems where workup was not differentiated from the broader diagnostic process.

One reason for this relative neglect of workup as a separate domain may be that many areas are so focussed and constrained that for a *noncritiquing* system they seem insufficiently complex. Certainly the mere listing of a recommended sequence of tests is computationally trivial. Once one takes the critiquing approach, however, the problems of workup become much more interesting, since the system must be able to adapt itself to *any* approach and still respond appropriately.

Patient workup has several advantages as a general area in which to explore the critiquing approach.

1. Compared to diagnosis and management, many areas of workup may indeed prove to be quite constrained. As a result, the knowledge bases required to implement useful systems may be relatively modest. If so, workup should prove a fertile ground for the development of practical systems in a reasonable time frame.
2. With the current nationwide concern about the high costs of health care, it becomes increasingly important to help the physician make his workup as rational and efficient as possible, especially since many workup procedures are expensive and time consuming.

6.2. The Workup of Pheochromocytoma

A pheochromocytoma is a catecholamine-secreting tumor, usually of the adrenal gland (Manger and Gifford 1977). It is one of the surgically treatable causes of hypertension but is very rare. Once the presence of a pheochromocytoma is suspected from a patient's history, the following modalities of workup are available:

Screening tests. The initial tests traditionally used in screening for pheochromocytoma have been urinary tests for catecholamines and their metabolites.
More recently a test for plasma catecholamines has also become available. A
positive (high) result from one of these tests is usually considered to be diagnostic of pheochromocytoma. Prior to performing these tests the physician
must make sure that the patient is not taking any drugs which might interfere
with the results.

Pharmacologic tests. If the initial tests are equivocal, or if they are negative but
the physician is still suspicious, then two types of pharmacologic test may be
performed. A provocative stimulation test, e.g., using glucagon, attempts to
elicit a catecholamine response. A clonidine suppression test suppresses
endogenous catecholamine secretion but does not alter the autonomous catecholamine release by a tumor.

Radiologic tests. Two radiologic tests are used in working up a patient for a
pheochromocytoma: CT scan and angiogram. These tests are used in two circumstances. (1) If the screening and pharmacologic tests are equivocal, or if
they are negative but the physician remains suspicious, then radiologic tests
can be used diagnostically to look for a tumor. (2) If the prior tests were positive, then radiologic tests can help the surgeon by pinpointing the tumor's location and anatomy.

Other tests. In addition to the tests mentioned above, there are other tests available, such as scintillation scanning. These tests, however, are not widely available and were not included in the present system.

6.3. Conflicting Expertise

An interesting issue that arose during the implementation of PHEO-ATTEND-
ING was that experts disagreed as to the best approach to working up a suspected
pheochromocytoma. In retrospect, this fact is not surprising. A new test has
recently become available: plasma catecholamines. Some experts, including
those who helped develop this new test, advocate its use. Other experts are skep-
tical of its superiority over previously available options. This disagreement also
reflects different experience, capabilities, and biases of the laboratories in the
hospitals where the experts practice.

Rather than try to ignore these differences in expert opinion, we decided to
make the problem of conflicting expertise a major focus of the present research.
To this end, two systems were initially implemented, embodying two conflicting
expert approaches. For simplicity, we will identify these as the BRAVO approach
and the BLACK approach.

1. *The BRAVO approach.* The system that implements this approach is modeled
   after a paper by Bravo (1983) on pheochromocytoma. (The system may not,
   however, fully represent the nuances of his practice.) This approach has the
   following features: (1) It advocates the use of plasma catecholamines in
   preference to urinary tests. (2) It then advocates either a clonidine suppression

test or a stimulation test (in a normotensive patient) if the plasma catecholamine result is equivocal. (3) Radiologic tests are used later, as described
above.

2. *The BLACK approach* (Black and Bursten 1984). This approach differs in
   several respects. (1) It recommends initial screening using urinary tests and
   suggests repeating them three times if they are low or equivocal, in an attempt
   to obtain a high (positive) value. (2) It does not use plasma catecholamines for
   screening but may use them later if urinary tests are equivocal. (3) It rarely
   uses a stimulation test but may use a clonidine suppression test after an
   equivocal plasma catecholamine result. (4) Radiologic tests are used later, as
   described above.

Both experts have reasoned arguments to justify their preferences. The two
approaches are outlined in overview in Figure 6.1.

The next section gives several examples illustrating how these two systems
react to different approaches to pheochromocytoma workup. *These two
approaches were implemented separately because we felt this would highlight the
differences in expert opinion explicitly* and would thereby be instructive to system
designers. The chapter later describes experiments in merging the two
approaches into a single critiquing advisor. Based on these experiments, some
observations are made as to how conflicting expertise may best be merged.

6.4. Examples

This section gives three examples illustrating how the BRAVO and BLACK
versions of PHEO-ATTENDING currently critique the workup of pheochromocytoma. Prior to each critique, the physician inputs a modest amount of information, including:

1. The patient's age, sex, and blood pressure
2. Any symptoms suggestive of pheochromocytoma
3. Any tests planned or ordered, together with any test results already obtained.

After each of the following examples, certain features of the critiques are
discussed.

Example 1. Here the physician has suggested initiating workup by ordering one
test: urinary catecholamines. In this example, both critiques start out with the
same introductory paragraph:

> Pheochromocytoma is a rare etiology of secondary hypertension. Its diagnosis is
> suspected on the basis of clinical manifestations and is confirmed by laboratory
> results. The presence of the sweating, tachycardia, headache triad is highly sugges-
> tive of a pheochromocytoma with a specificity and sensitivity of approximately
> 90%. The symptoms of palpitations, anxiety, and tremulousness are common in

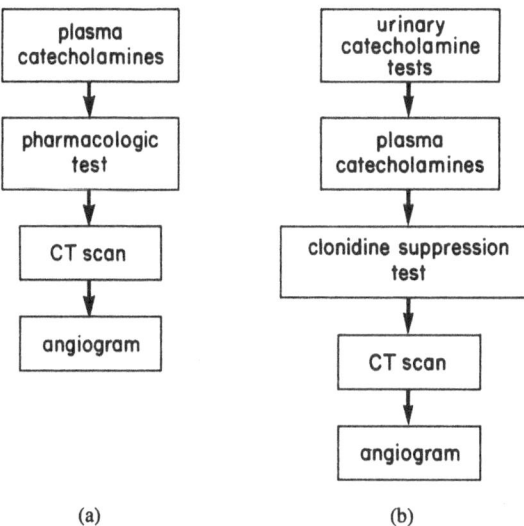

Figure 6.1. A rough overview of the BRAVO (a) and BLACK (b) approaches to pheochromocytoma workup. These diagrams indicate the preferred sequence for performing the tests. If a test result is positive, however, certain later tests may be omitted.

epinephrine-producing tumors. Interfering CNS drugs, such as methyldopa, levodopa, and dopamine, must be stopped before testing to obtain accurate laboratory results.

This introductory paragraph varies slightly in different critiques, depending on the clinical picture. For instance, the fourth sentence is included only if palpitations, anxiety, or tremulousness were reported. Similarly, if there were no symptoms suggestive of pheochromocytoma, the paragraph comments, "In the absence of symptoms, no testing is suggested unless a strong clinical suspicion is still present."

Since this introductory paragraph is included in each critique with only minor changes, it is omitted in the critiques that follow.

The BRAVO Critique

Measuring catecholamines or their metabolites is generally regarded as the first step in the workup when symptoms are present. If you do perform urinary tests, we would normally suggest you test for all three: urinary catecholamines, vanillylmandelic acid, and total urinary metanephrines. Urinary tests for catecholamines and their metabolites are the most widely used screening tests for pheochromocytoma. However, there is a fairly high incidence of false negatives, ranging in studies from 1 in 14 to 7 in 28.

On the other hand, in our experience plasma catecholamine values in patients with pheochromocytomas were greater than the highest values in patients without one. Thus we usually recommend plasma catecholamines be tested since we consider this to be the most reliable test for catecholamines.

The BLACK Critique

Measuring catecholamines or their metabolites is generally regarded as the first step in the workup when symptoms are present. We agree that urinary tests are the most appropriate first step. However, we normally advocate you test for all three: urinary catecholamines, vanillylmandelic acid, and total urinary metanephrines. The results from having ordered urinary tests will most likely resolve the diagnosis.

This example illustrates how the two approaches differ in their preferred approach to initial screening. The BLACK critique agrees with the proposed urinary test, while the BRAVO critique prefers plasma catecholamines and explains its reason for this preference. Both critiques indicate that if urinary tests are used, all three would be performed.

Several additional issues are discussed below:

1. Making sure tests are performed correctly. Just as important as the choice of an initial test is that it be performed correctly. The patient should be prepared properly prior to the test. Also, in performing urinary tests, creatinine is usually measured in the sample to help assure that a proper 24-hour specimen was obtained.

   One therefore *might* design the critique to include precise instructions as to how each test should be done. We have not taken this approach, feeling that a detailed set of instructions would detract from the impact of a concise critique. In a fully developed consultation system, however, we would suggest that these instructions be included online. After the critique, the system could ask if the physician wanted this further information.

2. The role of statistics. There are several different ways one might utilize statistical information in a critiquing system. The approach taken in PHEO-ATTENDING is to use statistical knowledge at the "surface" level of its analysis, to amplify and enhance the prose critique, as illustrated in the introductory paragraph and in the BRAVO critique. This is certainly not the only way to incorporate statistical knowledge. At one extreme, statistics could be used to drive a critiquing system's internal analysis. Alternatively, prior to implementing the system, one might perform a decision-theoretic analysis to determine exactly which alternatives the system should advocate.

   Whichever approach is taken, the judicious incorporation of statistics into the prose discussion may be helpful. Most clinical studies are in effect attempts to critique the different options that a physician faces in patient care. As a result, statistical knowledge gathered in such studies may be tailor made to fit into the prose discussion produced by a critiquing computer-advisor.

3. Truly puzzling cases. A system like PHEO-ATTENDING is designed to help the physician structure his workup efficiently, avoiding unnecessary tests, incorrect tests, or tests out of their appropriate sequence. There may be rare patients, however, whose presentation and workup confounds even the expert. For such a case, a human expert should provide the required consultation. A system like PHEO-ATTENDING is designed not for such truly

puzzling cases but to help the physician with the large majority of patients he encounters.

Example 2. Here, after an initial plasma catecholamine test has yielded an equivocal result, the physician has ordered a glucagon stimulation test in a hypertensive patient.

The BRAVO Critique

We feel that plasma catecholamines is the most reliable test for pheochromocytoma, having found that plasma catecholamine values in patients with pheochromocytomas were greater than the highest values in patients without one.

In the minority of cases when pheochromocytoma is suspected but cannot be excluded by catecholamine studies, a pharmacologic approach can be used. With equivocal plasma catecholamines, we normally advocate, in a hypertensive patient, a clonidine suppression test.

A glucagon stimulation test is a provocative pharmacologic approach for catecholamine tumors and is thus only considered for a normotensive patient when prior testing has not been helpful. Before attempting a glucagon stimulation test, you must be prepared for a possible catecholamine crisis. Thus it is preferable that your patient be in a hospital setting for this test. To avoid these risks you might want to consider a clonidine suppression test.

The BLACK Critique

We do not regard plasma catecholamine testing as an initial step since it is more prone to error, including false elevations even from venipuncture. We usually recommend urinary catecholamines be tested as part of our standard initial workup instead of plasma catecholamines.

In the minority of cases where pheochromocytoma is suspected but cannot be excluded by catecholamine studies, a pharmacologic approach may be used. Since a glucagon stimulation test is a provocative test, we consider it to have little role in the workup of pheochromocytoma. A clonidine suppression test is preferred when indicated.

In this example, the critiques differ regarding two separate parts of the workup. Again, there is the difference in preference for the initial screening. Also, although both critiques react negatively to the glucagon stimulation test, the BLACK critique is categorically negative, while the BRAVO critique is negative because the patient is hypertensive.

Example 3. Here, after all three initial urinary tests were low (negative), the physician has ordered a diagnostic CT scan.

The BRAVO Critique

Measuring catecholamines or their metabolites is generally regarded as the first step in the workup when symptoms are present. Urinary tests for catecholamines and their metabolites are the most widely used screening tests for pheochromocytoma. However, there is a fairly high incidence of false negatives, ranging in studies from 1 in 14 to 7 in 28.

On the other hand, in our experience plasma catecholamine values in patients with pheochromocytomas were greater than the highest values in patients without one. Thus we usually recommend plasma catecholamines be tested since we consider this to be the most reliable test for catecholamines.

In the presence of low urinary values, pheochromocytoma is normally considered to be ruled out. For reasons previously detailed, however, we are suspect of urinary tests. Thus if you are suspicious of your urinary values we normally recommend that a plasma catecholamine test be done.

After confirmation by laboratory chemistries, an abdominal CT scan is used to localize the tumor mass for subsequent surgical intervention. A CT scan may also be used diagnostically if previous workup is inconclusive. With the findings reported, a CT scan is not indicated at this time.

The BLACK Critique

Measuring catecholamines or their metabolites is generally regarded as the first step in the workup when symptoms are present. With low or equivocal urinary values, we suggest you test urinary catecholamines, total urinary metanephrines and vanillylmandelic acid a total of 3 times each for a possible high value. The literature suggests that this technique should catch virtually all patients with pheochromocytomas. If low urinary values are the result of 3 trials each, pheochromocytoma is considered ruled out.

After confirmation by laboratory chemistries, an abdominal CT scan is used to localize the tumor mass for subsequent surgical intervention. A CT scan may also be used diagnostically if previous workup is inconclusive. With the findings reported, a CT scan is not indicated at this time.

In this example, the physician suggested performing a CT scan. Both critiques agree that a CT scan is premature but disagree as to the best next step.

These three examples illustrate several points. First they demonstrate that pheochromocytoma workup is sufficiently complex that critiquing feedback may have value to the physician. At the same time, the domain is sufficiently constrained that the knowledge base required is not impractically large. The examples also illustrate concretely some of the issues of conflicting expertise that may arise in building an expert system that critiques patient workup.

6.5. Critiquing Workup: System-Design Considerations

This section describes PHEO-ATTENDING's internal design. As mentioned previously, it is anticipated that this design may be readily extended to other areas of patient workup. For this reason, the description has been couched in as general terms as possible. Although PHEO-ATTENDING was not implemented using ESSENTIAL-ATTENDING, it does embody ESSENTIAL-ATTENDING's basic design (see Chapter 7).

PHEO-ATTENDING's design has the following features:

1. "Expressive frames" are associated with each test and procedure. Each of these frames contains a list of *comments* that may be output in discussing the

use of that test. Each comment has an associated *condition* that indicates when it is to be output as part of the critique.

2. A "no-look-ahead" design. The conditions that determine which comments will be used in discussing a test look only at information about tests that precede or parallel the test in the workup sequence.

3. The individual comments. The individual comments are expressed using the PROSENET formalism (see Chapter 7), which allows their expression to be flexibly modified depending on the context in which they are used.

4. Prose generation. The system's final critique is assembled from the various comments using PROSENET, an approach developed to facilitate computer generation of polished prose.

The remainder of this section discusses each of these features in turn.

6.5.1. Expressive Frames

An "expressive frame" is associated with each test and procedure. It contains a list of comments that may be output when discussing that test, as illustrated in Figure 6.2. Associated with each comment is a condition, which indicates when that comment is to be included in the critique. These conditions test different types of information:

1. Aspects of the patient description. If a glucagon stimulation test is ordered for a patient who is hypertensive, a comment in the glucagon test frame indicates that this is hazardous.

2. The presence or absence of a previous test. If a CT scan is ordered without prior screening tests, a comment in the CT scan frame suggests that these should be ordered first and that a CT scan is not yet indicated.

3. The result of a previous test. If a urinary test is high (positive) but the physician has ordered a clonidine suppression test, a comment in the clonidine test frame indicates that the test is not necessary since a diagnosis has already been made, assuming that the high urinary result was not falsely positive.

4. Aspects of the current test. If not all the urinary tests are ordered, a comment in the urinary test frame discusses why all three should be performed.

5. Unconditional comments. Some comments are always output when discuss-

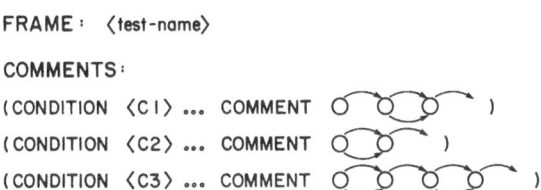

Figure 6.2. An example expressive frame. Such frames are associated with each test and procedure.

ing a test. Typically these are introductory comments that describe the test in general terms and discuss its clinical indications.

6.5.2. A "No-Look-Ahead" Design

One general restriction is placed on the design outlined above. As illustrated in Figure 6.3, the conditions associated with a frame examine information only about tests that precede or parallel that frame's test in the workup sequence. For instance, the catecholamine frame does not comment about the presence, absence, or results of a CT scan. This "no-look-ahead" restriction may not be necessary but seems to help structure the system in a logical, organized fashion.

6.5.3. The Individual Comments and Prose Generation

The individual prose comments are stored in a flexible form that allows their expression to vary depending on the context in which they are used. This flexibility is afforded by PROSENET, an approach developed to facilitate the generation of polished prose. Figure 6.4 shows two of PHEO-ATTENDING's comments expressed in PROSENET form. Once the system has decided which comments will be included in the critique, these are then assembled into paragraphs by the PROSENET prose generator described in Chapter 7.

6.6. Merging Conflicting Expertise

The previous sections have described how two conflicting expert approaches to the workup of pheochromocytoma were implemented in separate critiquing systems. This section describes experiments in merging the approaches into a single critiquing system.

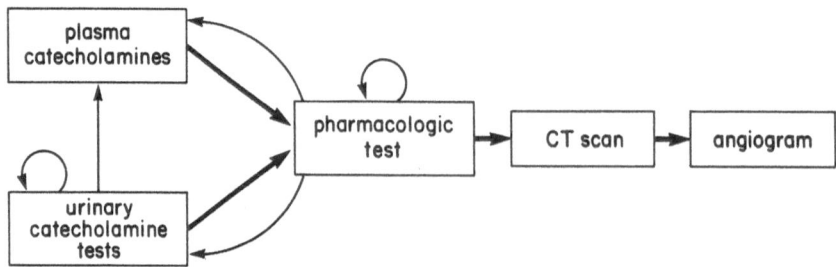

Figure 6.3. This figure illustrates that the conditions in each test frame do not look at later tests in the workup sequence. In this figure, the broad arrows indicate the normal sequence of workup. The narrow arrows illustrate that the comments in a test frame looks only at information about previous or parallel tests.

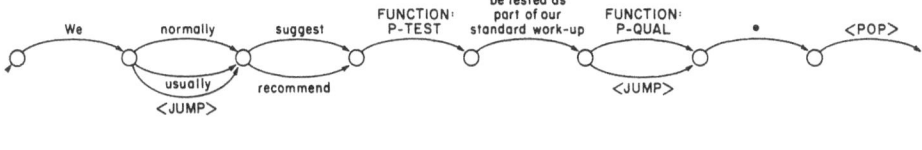

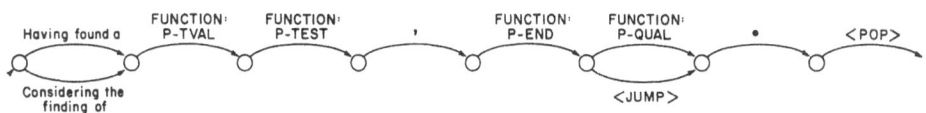

Figure 6.4. Two example comments expressed as PROSENET networks. These are simplified versions of networks used by PHEO-ATTENDING in constructing its critique. Most arcs in a network have an associated prose fragment or prose-generating function (e.g., P-TEST, P-QUAL). When an arc is traversed, the associated prose is output as part of the critique.

As illustrated in Figure 6.5, the two approaches were merged by combining corresponding test frames. A combined frame for a test was constructed from the comments of the two conflicting frames, making any modifications necessary to help the merged critique flow smoothly. In this way, the process of merging the two approaches is largely reduced to the subproblem of merging individual test frames.

It would be interesting to explore ways to help automate this process. Perhaps the merging could be performed by the system itself. Alternatively, the system might be designed to assist the designer to build and then merge the frames. Such automation was not attempted in the present research.

There are, however, different strategies that can be employed in the merging process described above. To explore these strategies, frames from the BLACK and BRAVO approaches were merged in different ways. In these experiments, only the urinary and plasma catecholamine frames were used, since the two

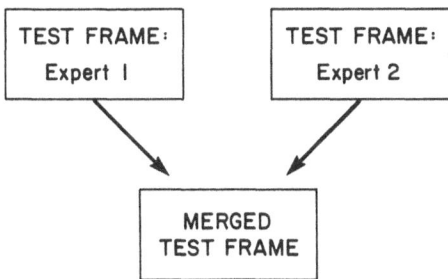

Figure 6.5. Comments from the two conflicting expert approaches are merged experimentally frame by frame.

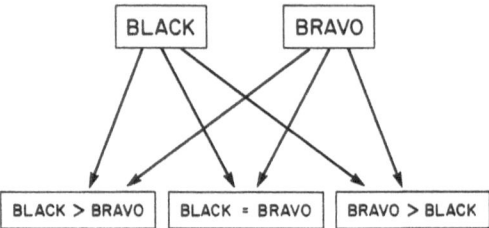

Figure 6.6. The two conflicting expert approaches can be combined using different merging strategies.

approaches differed most markedly in this area. To construct a complete combined system, a more comprehensive merging of the two approaches would have to be performed.

For these experiments, as illustrated in Figure 6.6, three merged systems were defined (BLACK = BRAVO, BLACK > BRAVO, and BRAVO > BLACK), each reflecting a different merging strategy. Of these, two were implemented. The following example critiques illustrate the character of these merged systems.

The BLACK = BRAVO Critique. Here a physician has ordered all three urinary screening tests.

> Measuring catecholamines or their metabolites is generally regarded as the first step in the workup when symptoms are present. Black agrees that urinary tests are the most appropriate first step. The results from having ordered urinary tests will most likely resolve the diagnosis.
> On the other hand, Bravo argues that urinary tests involve a fairly high incidence of false negatives, ranging in studies from 1 in 14 to 7 in 28. In his experience plasma catecholamine values in patients with pheochromocytomas were greater than the highest values in patients without one. Thus he usually recommends plasma catecholamines be tested since he considers this to be the most reliable test for catecholamines.

This merged critique is designed to maintain neutrality, exhibiting no inherent preference for either approach. Thus, if the physician orders plasma catecholamines, the system comments that Bravo agrees but that Black disagrees and indicates in each case why. Similarly, if the physician orders urinary tests, as illustrated above, the system indicates why Black agrees and why Bravo disagrees.

Notice that the comments in this critique are slightly edited versions of comments from the BRAVO and BLACK critiques in example 1. The expressive frame that generates this merged critique was assembled from the two "parent" frames in a straightforward fashion. The first person ("we") was changed to the third person, and other minor changes in wording were made to let the merged critique flow smoothly.

The BLACK > BRAVO Critique. Here a physician has ordered an initial plasma catecholamine test.

> Measuring catecholamines or their metabolites is generally regarded as the first step in the workup when symptoms are present. We do not regard plasma catecholamine testing as an initial step since it is more prone to error, including false elevations even from venipuncture. We usually recommend urinary catecholamines be tested as part of our standard initial workup instead of plasma catecholamines.
>
> On the other hand, Bravo argues that urinary tests involve a fairly high incidence of false negatives, ranging in studies from 1 in 14 to 7 in 28. In his experience plasma catecholamine values in patients with pheochromocytomas were greater than the highest values in patients without one. Thus he agrees with your decision that plasma catecholamines be tested since he considers this to be the most reliable test for catecholamines.

This merged critique is designed to have a preference for the BLACK approach. Thus, if the physician orders urinary tests, the system merely indicates why that approach is preferred and does not mention Bravo's approach. If, however, plasma catecholamines are ordered, as illustrated above, this critique discusses why it recommends against this but at the same time discusses why Bravo advocates the test.

The BRAVO > BLACK Critique. A third merged system, with the preference for the BRAVO approach, was not implemented. Since this was the converse of BLACK > BRAVO, no new design issues would have been explored.

In this way, two basic differences in *merging strategy* were explored:

1. A neutral strategy. This strategy yields a system that is not biased and that always give both sides of any argument.
2. A "system-preference" strategy. This approach yields a system that sides with one expert and therefore has its own preferred approach. It discusses the other expert approach only if the physician himself chooses it.

Neither of these strategies is right or wrong. In implementing a critiquing system that combines conflicting expertise, one might choose either strategy depending on the nature of the domain and on the type of system desired.

The following observations may help in making this decision:

1. The neutral approach tends to be wordier, since it always gives both sides of the issue. This may lessen the system's impact. Also, the approach might become especially unwieldy if more than two expert approaches were incorporated into a system. Here, the neutral approach might prove impractical.
2. The neutral approach might be valuable if the system designers advocated a new or unconventional approach. Here, no matter which approach the physician chooses, the designers may want to mention both their preference and the more conventional approach, if only for comparison and instruction.

3. The neutral approach might be difficult if there were major differences in workup sequence. In our experiments, we looked only at merging individual test frames and did not look at differences in sequence. One advantage of the system-preference strategy is that the preferred approach gives a natural order to the system's critique. Finding a neutral order for the critique might prove difficult if there were major differences in the experts' workup sequence.
4. The system-preference approach allows a hospital laboratory to express its biases as to how workup should be done. A laboratory frequently may wish to guide physicians in certain directions. One could still allow the physician to see how a conflicting expert would critique his workup.

In summary, a critiquing system should be able to incorporate conflicting expert points of view. If it does not, it will be significantly limited in its knowledge and in the maturity of its advice. Different domains and specific laboratory biases may make different strategies more appropriate when merging conflicting expert opinions.

The goal of our experiments in merging the BLACK and BRAVO approaches is to demonstrate that this may be done in large part by merging frames from the separate approaches and to explore the relative merits of different strategies that might be employed.

6.7. PHEO-ATTENDING: Summary

PHEO-ATTENDING was developed as a research vehicle to explore the system-design issues involved in critiquing patient workup. The system has not been subjected to clinical evaluation. Indeed, because of the rarity of pheochromocytomas, it would be difficult to do so. Nevertheless, the system does illustrate a number of design concepts that may be extended to other workup domains.

In addition, the system demonstrates that test and procedure "frames," developed separately to reflect conflicting expert approaches, can be merged into a single critiquing system. Because of the constant evolution of medical practice, the problem of conflicting expertise will be encountered frequently. Experts are subject to the same practice variation that pervades medicine as a whole.

The critiquing approach itself is designed to accommodate practice variation on the part of the user. *To maximize the impact of its advice, however, a critiquing system must accommodate practice variation both on the part of experts and on the part of practitioners*. The system must mediate between the expert and the user in a domain where nothing is necessarily fixed or absolute. The present research explores how this might be accomplished in areas of workup.

Patient workup is a central area of medicine. It is important that workup be performed efficiently, and that a physician be able to obtain feedback to help optimize his practice. The critiquing approach is a vehicle that allows the computer to provide this feedback in a natural way, tailored to the physician's daily patient care.

Chapter 7

ESSENTIAL-ATTENDING: Building Expert Critiquing Systems*

ESSENTIAL-ATTENDING (E-ATTENDING) is a "system-building system" designed to assist in the implementation of an expert critiquing system. E-ATTENDING has been refined in the process of implementing several developmental critiquing systems, including VQ-ATTENDING and HT-ATTENDING. It is currently being applied in areas of medical management, patient workup, and differential diagnosis.

E-ATTENDING is designed to include the domain-independent components required to help implement a class of critiquing systems, in a subset of possible critiquing domains. It may also be augmented in various ways by interested users to accommodate critiquing domains with more complexity.

7.1. Expert System-Building Systems

E-ATTENDING can be perceived (1) as a set of domain-independent tools, (2) as a critiquing "system-building system," or even (3) as a special-purpose programming language. Such a system offers several advantages:

1. The most obvious advantage is that certain programs required to implement a critiquing system are already included in E-ATTENDING.
2. A more subtle advantage is that a domain-independent system forces an existing design on the critiquing system builder. Several central design decisions have already been made. Once the system builder understands E-ATTENDING's design constraints, much of his overall design is already done. Since this initial formulation is often a difficult and critical process, a system-building system can be very helpful as long as its design is appropriate and is sufficiently general to adapt to new domains.

---

*Adapted from P.L. Miller, Building an Expert Critiquing System: ESSENTIAL-ATTENDING. In *Methods of Information in Medicine*, v. 25, no. 2, 1986. F.K. Schattauer Verlag. Reprinted by permission.

Several previous expert system-building systems have been constucted in medicine. E-MYCIN (van Melle 1979) embodies the domain-independent components of MYCIN (Shortliffe 1976), a system originally implemented for infectious disease. In addition, EXPERT (Weiss and Kulikowski 1979) and KMS (Reggia and Pericone 1981) were both designed to help build expert medical consultation systems. No previous system-building system, however, has been designed to facilitate the implementation of a *critiquing* system.

7.2. The Motivation Behind ESSENTIAL-ATTENDING

The primary motivation behind the development of E-ATTENDING is to help others implement expert critiquing advisors. To this end, E-ATTENDING demonstrates a possible system design. It is anticipated that the current E-ATTENDING can be used to build a class of critiquing systems in selected domains, as discussed in Section 7.10. The remainder of this section discusses a number of issues to help place the current E-ATTENDING into perspective.

1. E-ATTENDING is quite simple. Compared to most software systems, E-ATTENDING is quite simple. As shown in the Appendices, the entire E-ATTENDING system comprises only 6 to 7 pages of LISP code. There are many ways in which this simple E-ATTENDING could be augmented, as discussed in Section 7.10. The rationale for keeping E-ATTENDING simple is: (1) in a number of medically interesting domains, the creation of critiquing advisors need not be computationally complex, and the current E-ATTENDING can be used unchanged. (2) If E-ATTENDING is kept simple, a user can become familiar with the whole system in detail. He can then augment it in a number of interesting ways, adapting it to the requirements of different critiquing domains.

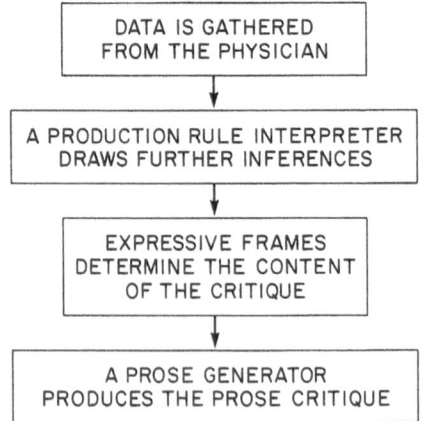

Figure 7.1. Schematic overview of a critiquing system built using ESSENTIAL-ATTENDING.

2. There is nothing inherently medical about E-ATTENDING. Although E-ATTENDING evolved from the implementation of several medical critiquing systems, it can certainly be applied to nonmedical domains as well.
3. There is nothing inherently "critiquing" about E-ATTENDING. The current version of E-ATTENDING is not limited to building critiquing systems. The design could be used to help any computer system produce complex prose analysis.

Despite these disclaimers, however, the emphasis in developing E-ATTENDING is to facilitate the implementation of expert critiquing systems in medicine.

In summary, the current E-ATTENDING system is designed for a subset of possible critiquing domains. The system attempts (1) to "extract the essence" from the critiquing systems already implemented, (2) to structure the process of building a critiquing system, and (3) to provide a base from which others may explore how critiquing design concepts can be augmented and refined.

7.3. E-ATTENDING System Design: Overview

A critiquing system built using E-ATTENDING has four components, as shown in Figure 7.1.

1. First, *data* are gathered from the physician. These data include information about the patient and about the physician's planned approach to some medical problem.
2. Next, these data are passed to a *production rule interpreter*, which uses the data to draw further inferences that will be used in the system's later analysis.
3. The initial data and any inferred conclusions then serve as input to modules that manipulate *expressive frames*. These frames contain information about the various alternatives the physician faces: e.g., different drugs and management techniques or different tests and procedures. Based on the processing of these expressive frames, the system determines the content to be included in the critique. This "content" may be a set of comments to be incorporated into prose paragraphs or may be information from which a set of comments is to be constructed.
4. Finally, this content to be included in the critique is input to a *prose generator*, which sequences the material and assembles it into a set of prose paragraphs critiquing the proposed approach.

The first stage of this process, which gathers information from the physician, is hand coded by the system builder. Domains of medicine are so dissimilar that it is difficult to impose a rigid structure on the collection of this data. The interaction should be as smoothly tailored to the domain as possible. Therefore a general data collection interface is not included in E-ATTENDING.

As a result, E-ATTENDING itself consists of three domain-independent components:

1. A production rule interpreter

2. Routines that manipulate the expressive frames
3. A prose-generating system.

To use E-ATTENDING, a system builder must describe his domain by specifying appropriate production rules, expressive frames, and prose-generating logic. In addition, he must write a top-level program that gathers information from the physician and coordinates the overall analysis. He may also choose to write other programs to augment the system in various ways. (Since E-ATTENDING is written in LISP, the construction of critiquing systems using E-ATTENDING is done in a LISP programming environment.)

7.4. A Simple Example

This section shows how a simple example system is built using E-ATTENDING and illustrates in overview how the system components fit together. (These system components are described in more detail in Sections 7.5, 7.6, and 7.7.) The example system is a simplified portion of HT-ATTENDING, in the domain of essential hypertension, critiquing the proposed use of a thiazide diuretic.

7.4.1. Production Rules

The example system has a single production rule:

```
(SETQ THIAZIDE RULES '(

(CONTRA_THIA RULE
            (IF (AND (SAME AGENT THIAZIDE)
                    (SAME HX CRF))
            (THEN (CONTRAINDICATION THIAZIDE)))
))
```

An English translation of this rule is: "If a thiazide diuretic is proposed and there is a history of chronic renal failure, then there is a relative contraindication to the thiazide diuretic." The conclusion made by this rule is tested in the expressive frame and prose-generating logic shown below.

7.4.2. Expressive Frames

The example system has a single expressive frame.

```
(DEFFRAME 'THIACRITIQUE '(

(IF T COMMENT $THIA_INTRO SEQUENCE 1)

(IF (SAME HX CRF) COMMENT $THIA_CRF SEQUENCE 2)

(IF (AND (SAME HX CRF)(SAME AGENT METOLAZONE))
            COMMENT $METOL_CRF SEQUENCE2.2)
```

(IF (SAME HX DIABETES) COMMENT $THIA_DIAB SEQUENCE 2.5)

(IF (AND (NOT (SAME CONTRAINDICATION THIAZIDE))
       (SAME AGENT HYDROCHLOROTHIAZIDE)) COMMENT $HCTZ
       SEQUENCE 3)

(IF T COMMENT $THIA_ALTS SEQUENCE 4)

) 'GENCOMMENTS)

This expressive frame lists six "comment frames," each indicating a comment that might critique the use of a thiazide diuretic. Each comment frame has three parts: (1) a condition, (2) a pointer to the prose comment itself, and (3) other information, which in this case is an indicator to help sequence the critique.

For example, the second comment frame indicates that "if the patient has a history (HX) of chronic renal failure (CRF)," then the comment $THIA_CRF should be included ("in sequence position 2") in the critique.

7.4.3. The PROSENET Comments

The prose comments that correspond to the six comment frames of this expressive frame are shown below. These comments are expressed using PROSENET, in the augmented transition network (ATN) formalism. As described in Section 7.6, PROSENET allows the phrasing of each comment to vary flexibly depending on its context.

(DEFPROSE '(

($THIA_INTRO ((*para a thiazide diuretic or similar acting agent
              is the drug of choice for most patients *period) $ POPTT T))

($POPTT (*pop T T))

($THIA_CRF ((however *comma if this patient has significant renal insufficiency
           *openparen glomerular filtration rate less than 50 ml per minute
           *closeparen *comma then a loop diuretic like furosemide would
           generally be considered most appropriate *period) $POPTT T))

($HCTZ ((hydrochlorothiazide is a commonly used thiazide diuretic *period)
        $POPTT T))

($THIA_ALTS ((*para if you do use a thiazide diuretic *comma)
        $TA1O (SAME CONTRAINDICATION THIAZIDE))
        (*jump $TA1O T))

($TA1O ((chlorthalidone or metolazone) $TA2O
            (NOT (OR (SAME AGENT METOLAZONE)(SAME AGENT
            CHLORTHALIDONE))))
       (chlorthaladone $TA45 (SAME AGENT CHLORTHALIDONE))
       (metolazone $TA45 T))

($TA2O ((are alternative agents worth considering since they are long-acting and
         can be given once a day *period) $POPTT T))

```
($TA45 ((has the advantage that it is long-acting and can be given once a day
         *period) $POPTT T))

($METOL_CRF ((some patients with renal insufficiency *comma however
              *comma may respond to metolazone *period) $POPTT T))

($THIA_DIAB ((in a patient with diabetes *comma one must remember that a
              thiazide diuretic may increase glucose intolerance *period)
              $POPTT T))
))
```

Each PROSENET network consists of a state name (e.g., $THIA_INTRO, $THIA_CRF) followed by one or more arcs. Each arc consists of (1) a prose fragment (or some prose-generating action), (2) a "destination state," and (3) a condition. In processing such an arc, if the condition is *true*, the prose fragment is output and processing continues at the "destination state." If the condition is *not true*, processing continues with the next arc of the current state. Processing ends when a "*pop" arc is traversed.

7.4.4. Operating the Example System

To operate the example system, initial data describing the patient and the proposed plan must be "asserted":

```
(SETQ FACTS NIL)
(DEFFACT '(HX CRF))
(DEFFACT '(AGENT THIAZIDE))
(DEFFACT '(AGENT METOLAZONE))
```

This sequence of LISP statements (1) initializes the set of "facts" known about the world to NIL, (2) asserts that the patient has a history of CRF, and (3) and (4) asserts that a physician has proposed treatment with metolazone, a thiazide agent. Once this initial data is asserted, the production rules and prose generator are invoked.

```
(MAPC 'GENRULE THIAZIDE_RULES)
(GENFRAME 'THIACRITIQUE)
```

This activates (1) the rule interpreter and (2) the expressive frame analysis (which in turn calls the prose generator). As a result, the following prose is produced:

> A thiazide diuretic or similar acting agent is the drug of choice for most patients. However, if this patient has significant renal insufficiency (glomerular filtration rate less than 50 ml per minute), then a loop diuretic like furosemide would generally be considered most appropriate. Some patients with renal insufficiency, however, may respond to metolazone.
>
> If you do use a thiazide diuretic, metolazone has the advantage that it is long-acting and can be given once a day.

Below, a different patient and plan are described:

```
(SETQ FACTS NIL)
(DEFFACT '(HX DIABETES))
(DEFFACT '(AGENT THIAZIDE))
(DEFFACT '(AGENT HYDROCHLOROTHIAZIDE))
(MAPC 'GENRULE THIAZIDE_RULES)
(GENFRAME 'THIACRITIQUE)
```

For this input, the system produces the following critique:

> A thiazide diuretic or similar acting agent is the drug of choice for most patients. Hydrochlorothiazide is a commonly used thiazide diuretic. Chlorthalidone or metolazone are alternative agents worth considering since they are long-acting and can be given once a day. In a patient with diabetes, one must remember that a thiazide diuretic may increase glucose intolerance.

The next three sections describe the components of E-ATTENDING in detail.

7.5. E-ATTENDING Design: Production Rules

Production rules have been used extensively in expert system research to structure domain knowledge, including in MYCIN (Shortliffe 1976), EMYCIN (van Melle 1979), EXPERT (Weiss and Kulikowski 1979), and KMS (Reggia and Pericone 1981).

A production rule is a simple computational construct. It has two parts, an "IF-clause" and a "THEN-clause." The IF-clause may involve a single test or a combination of tests. The THEN-clause contains one or more conclusions, which are asserted if the IF-clause is true. These conclusions are considered to be "facts" about the "world" and are stored on a FACTS list. In this way, the "facts" may be tested by later rules.

Production rules are individual "chunks" of knowledge about the domain. There are several advantages of storing knowledge in this form (rather than using a more conventional procedural language):

1. The knowledge may be more easily inspected, modified, and augmented by the system designer and by domain experts.
2. The system itself may be better able to inspect the knowledge: (1) to test the knowledge for consistency and for completeness, (2) to use the knowledge to explain its inferencing steps, and (3) to assist in modifying and updating the knowledge.

7.5.1. The FACTS List

E-ATTENDING stores facts using a FACTS list. Any facts asserted by a rule's THEN-clause are placed on this list. Other facts may be asserted using the routine DEFFACT (for instance, the initial data gathered from the physician).

Each fact has three parts (1) a name, (2) an attribute, and (3) a value, stored in sequence (NAME ATTRIBUTE VALUE), for instance:

    (GOAL1 PRIORITY HIGH)

Two element facts are also legal, e.g., (GOAL1 ACTIVE). In this case, E-ATTENDING adds a third element (T), and stores the fact as (GOAL1 ACTIVE T). (If a LISP expression is used instead of a LISP atom as the NAME or ATTRIBUTE element, then that expression is evaluated, and is expected to return an atom, before the fact is stored.)

7.5.2. The IF-Clause

The IF-clause of a rule may contain a single test or a combination of tests. Each test is a LISP function, and the whole IF-clause is evaluated directly by the LISP interpreter. As a result, arbitrarily complex LISP programs may be placed in an IF-clause of a rule. In practice it is desirable to keep the tests simple and easily understood, for example:

```
(RULE MAINT_PO2
            (IF (OR (SAME PO2 ADEQUATE)
                    (SAME PO2 GOOD)
                    (SAME PO2 EXCELLENT)))
            (THEN (MAINT_PO2_GOAL ACTIVE)))

(RULE REDUCE_O2_TOX
            (IF (AND (NOT (SAME PO2 POOR))
                     (GREATERP FIO2CURR .6)))
            (THEN (REDUCE_O2_TOX_GOAL ACTIVE)))

(RULE RED_OX_SUPPORT
            (IF (AND (OR (SAME PO2 GOOD)
                         (SAME PO2 EXCELLENT))
                     (OR (GREATERP FIO2CURR .4)
                         (GREATERP PEEPCURR 5)))
            (THEN (REDUCE_O2_SUPPORT_GOAL ACTIVE)))
```

7.5.2.1. Builtin Functions that Test "Facts". E-ATTENDING has two builtin functions that test facts on the FACTS list: SAME and TEST.

1. *SAME* tests the FACTS list for a specific fact. Thus (SAME GOAL1 PRIORITY HIGH) will be true if the fact (GOAL1 PRIORITY HIGH) has been asserted, either by the THEN-clause of a rule, or by the function DEFFACT.
2. *TEST* is a more general function. It applies a specific LISP function to the "value" of a fact, e.g. (TEST *LESS PO2 CURRENT 60). This test searches for a stored fact of the form (PO2 CURRENT X), where X is any value, and returns the result of (*LESS X 60). Thus (SAME GOAL1 PRIORITY HIGH) is equivalent to (TEST EQUAL GOAL1 PRIORITY HIGH).

7.5.3. The THEN-Clause

The THEN-clause of a rule is very straightforward. It contains one or more facts to be asserted if the IF-clause is true.

7.5.4. Summary: E-ATTENDING's Production Rules

E-ATTENDING's production rule capability is quite simple. It allows inferences to be made in a rule-based fashion similar to that of many expert systems. The system designer must decide how much logic he wants to express using production rules, and how much he wants to express as conventional LISP procedures.

There are a variety of features that could be added to make the rule interpreter more sophisticated, such as (1) a trace facility; (2) an "explanation capability," which could use rules to help explain conclusions that had been asserted; and (3) knowledge inspection, knowledge verification, and knowledge acquisition aids, etc. As mentioned previously, the current system has been deliberately kept simple so that it can be understood and modified if desired.

7.6. E-ATTENDING Design: PROSENET

PROSENET is an approach developed to facilitate the generation of *polished prose* by a computer. The approach uses the augmented transition network (ATN) formalism (Woods 1970) which has been widely applied in areas of natural language processing. PROSENET's use of the ATN differs in that *prose fragments* and *prose-generating actions* are stored along ATN arcs. To help understand PROSENET's approach, it is helpful to consider other approaches taken to natural-language (English prose) generation by machine.

A more ad hoc approach. Most programs that generate English text do not have special capabilities to coordinate text generation. Strings of text are embedded in the program to be output where appropriate. When a system's analysis is straightforward, this approach may be satisfactory. When a system designer wants the prose output to adapt itself flexibly to the situation being described, however, this approach may be very limiting. The logic that generates the prose may become awkwardly intertwined with the logic that performs the underlying analysis.

A semantic approach. At the other extreme is a semantic approach. Here, the prose generation starts from an underlying representation of meaning. The system then proceeds to build phrases, clauses, and finally whole sentences. This approach represents a major ongoing research effort in its own right. It therefore may be impractical for a project whose emphasis is on medicine rather than natural-language research.

Any author knows how difficult it is to write polished prose. It is even more difficult to create a computer system that produces polished prose, particularly if many variables must be anticipated which influence what is to be said. The PROSENET approach has a number of advantages as a practical method of developing computer systems that produce polished prose.

Compared to the more ad hoc approach, PROSENET structures its prose comments using a powerful formalism (the ATN) developed to facilitate natural-language processing. Using this formalism, PROSENET offers a clean separation of the logic that performs the underlying analysis from the logic that creates the prose critique.

Compared to the semantic approach, PROSENET avoids the complex linguistic problems involved in sophisticated natural-language generation. By allowing prose fragments to be stored along ATN arcs, PROSENET takes a pragmatic approach, which nevertheless gives a system designer great flexibility in refining the prose output of an expert system.

7.6.1. Using PROSENET

Section 7.4 showed several example PROSENET networks. A sample PROSE-NET ATN in schematic form is shown in Figure 7.2. An ATN consists of *states* connected by *arcs*. Processing of the ATN starts at the *initial state* of a network, and follows arcs from one state to another, thereby traversing a *path* through the network. The path ends when a *"pop"* arc is traversed.

The exact path taken is controlled by *action routines*. An action routine is associated with each arc, and is a test performed prior to traversing that arc. If the test is *true*, then the arc is *active* and is traversed. If the test is *false*, then the arc is *inactive*, and some other arc must be taken. In this way, contextual information can be tested and can control the prose which is generated.

As illustrated in Figure 7.2, most arcs have an associated *prose fragment* or a *prose-generating action*. Whenever an arc is traversed, the associated prose fragment is output as part of the critique. The example ATN of Figure 7.2 is represented internally in PROSENET as follows:

```
($THIA_ALTS ((*para if you do use a thiazide diuretic *comma)
             $TA1O (SAME CONTRAINDICATION THIAZIDE))
          (*jump $TA1O T))

($TA1O ((chlorthaladone or metolazone) $TA2O
             (NOT (OR (SAME AGENT METOLAZONE)(SAME AGENT
             CHLORTHALIDONE))))
          (chlorthaladone $TA45 (SAME AGENT CHLORTHALIDONE))
          (metolazone $TA45 (SAME AGENT METOLAZONE)))

($TA2O ((are alternative agents worth considering since they are long-acting and
             can be given once a day *period) $POPTT T))

($TA45 ((has the advantage that it is long-acting and can be given once a day
             *period) $POPTT T))

($POPTT (*pop T T))
```

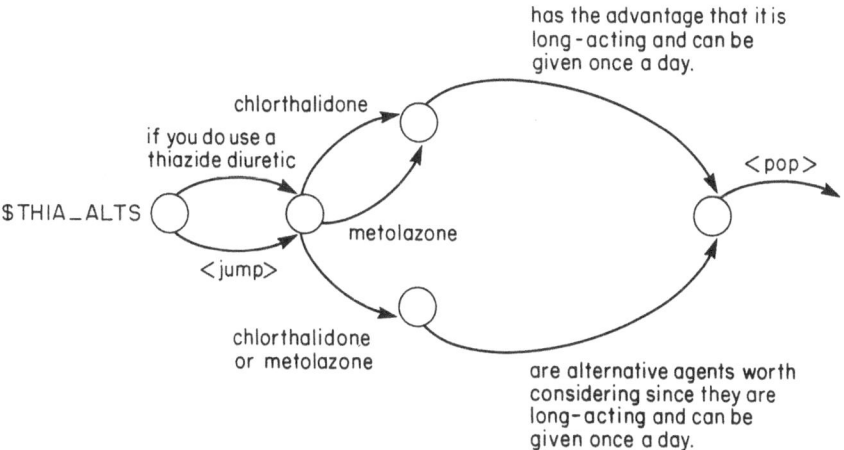

Figure 7.2. A sample PROSENET network.

Notice that each *state* is represented by a name ($THIA_ALTS, $TA1O, $TA2O, $TA45, $POPTT). Each arc is represented by three elements:

1. An *action*, which may be a *prose fragment* or some *prose-generating action*, as described below,
2. A *destination state*, which corresponds to the state pointed to by the arc in the schematic diagram.
3. An *action routine*, which may be any LISP expression.

For instance, in the first arc of the first state above,

1. The action is the prose fragment "*para if you do use a thiazide diuretic *comma."
2. The destination state is $TA1O.
3. The action routine is (SAME CONTRAINDICATION THIAZIDE).

In processing an arc, if its condition is true, the arc's action is performed (e.g., its prose fragment is output) and processing continues at the "destination state." If the condition is not true, processing continues with the next arc of the current state.

7.6.2. Actions That May Be Associated with an Arc

An arc may have eight different types of actions.

1. A *prose fragment* may be a single word or a list of words. Special keywords are used to specify punctuation, such as *comma, *period, *colon, *openparen, *para, *newline. Examples:

```
(*para $NS T)
(first $NS T)
((*comma however *comma) $NS T)
```

2. A *function arc* specifies a LISP expression, which might, for instance, test an item being discussed and generate prose describing that item. Examples:

```
((*function (PRINTGOAL)) $NS T)
((*function (PRINTGOAL (CAR GOALS))) $NS T)
```

3. An *option arc* specifies a list of arcs. One of these arcs is selected at random if the *option arc itself is traversed. Once one arc is used, it is temporarily removed from the list. Once all the arcs are used (by subsequent usage of the *option arc), the list is reinitialized. An *option arc allows a critiquing system to vary its phrasing and therefore avoid sounding monotonous. Example:

```
((*option ((from the standpoint of) $NS T)

          ((from the perspective of) $NS T)

          ((looking next at) $NS T)

          ) T T)
```

4. A *sequence arc* is similar to an *option arc in that it indicates a list of arcs, one of which is selected if the *sequence arc is traversed. Instead of being selected at random, however, the arcs are used in sequence. Therefore each time the *sequence arc is traversed, successive arcs are taken from the list. When only one arc remains, it is reused. A nametag must be included in the *sequence arc (e.g., SEQ1, in the example below). Example:

```
((*sequence SEQ1 ((first *comma) $NS T)

                 ((second *comma) $NS T)

                 ((third *comma) $NS T)

                 ((in addition *comma) $NS T)

                 ) T T)
```

5. An *endsequence arc* reinitializes the *sequence arc. The nametag of the *sequence arc must be specified. Example:

```
((*endsequence SEQ1) $NS T)
```

6. A *jump arc* is a "no-op." When a *jump arc is traversed, processing transfers to the destination state, but no other action is taken, and no prose is generated. Example:

```
(*jump $NS T)
```

7. A *push arc* transfers processing to a subnetwork and is equivalent to a prose-generating subroutine. Processing in the current network is suspended. Pro-

cessing starts at the indicated subnetwork and continues there until a *pop arc is traversed. At that time the suspended "higher level" network is reactivated. Example:

```
((*push $LOWERNET) $NS T)
```

8. A **pop arc* terminates processing in the current PROSENET network. Example:

```
(*pop T T)
```

The current implementation of PROSENET allows considerable flexibility in generating a prose critique but can be augmented in many ways.

7.7. E-ATTENDING Design: Expressive Frames

Expressive frames are used by E-ATTENDING to coordinate the construction of its prose critique. Expressive frames interface between the inferences made by production rules and the PROSENET comments. As a result, they are described now, after the other two system components have been discussed.

7.7.1. Why Expressive Frames?

To understand the role of expressive frames, one must consider the problem of constructing a complex prose critique without them. Suppose that there are 20 comments that *might* be included in part of a critique (depending on the particular patient and plan), but that only five to seven of the comments are typically used at any time.

One approach would be to string all 20 comments together using only PROSE-NET itself. Arcs would proceed directly from one comment to the next. At the start of each comment, appropriate tests would determine whether the system should output that comment or skip to the next comment.

This approach is very inflexible. All the processing takes place at one level. The system is not allowed a global overview view of the critique it is constructing. As a result, it is very difficult to coordinate the insertion of connective words and phrases between the various comments to help the train of thought flow more smoothly. Also, it would be extremely difficult to let the sequence of the comments vary depending on which were selected.

Expressive frames allow E-ATTENDING to avoid these problems. In an expressive frame each comment is represented by a "comment frame." This comment frame includes a test that indicates when that comment should be included in the critique. The comment-frame may also contain other information about the comment.

Using expressive frames, therefore, the 20 comments mentioned above are represented in an expressive frame by 20 comment frames. The first step in generating prose is to evaluate each comment's condition. This yields the subset of five to seven chosen comments. These can then be inspected as a group and sequenced in any way the system chooses. Also, if appropriate, additional connective words and phrases can be inserted between them. (Section 7.8 describes how a paragraph-level PROSENET network can coordinate this process.) In this way, the process of assembling a selected set of comments into an articulate critique is greatly facilitated.

7.7.2. The Expressive Frame

An example expressive frame was shown in Section 7.4. A portion of that frame is shown below:

```
(DEFFRAME 'THIACRITIQUE '(

(IF T COMMENT $THIA_INTRO SEQUENCE 1)

(IF (SAME HX CRF) COMMENT $THIA_CRF SEQUENCE 2)

(IF (AND (SAME HX CRF)(SAME AGENT METOLAZONE))
        COMMENT $METOL_CRF SEQUENCE 2.2)
.
.


) 'GENCOMMENTS)
```

Each expressive frame has three parts: (1) a name, (2) a set of comment frames, and (3) a function that coordinates the processing of the frame. The name (THIACRITIQUE) is self-explanatory. The other two components of the expressive frame are described below.

7.7.3. Comment Frames

As mentioned above, an expressive frame contains an arbitrary number of comment frames, as was illustrated in Section 7.4. Examples of comment frames are:

```
(IF (AND (SAME HX CRF)(SAME AGENT METOLAZONE))
        COMMENT $METOL_CRF SEQUENCE 2.2)

(IF (SAME ACHIEVE_PO2_GOAL ACTIVE) GOAL ACHIEVE_PO2_GOAL
        COMMENT (to achieve an adequate paO2))

(IF (SAME AGENT DIURETIC) FRAME DIURETIC SEQUENCE 1)
```

A comment frame has two required components: (1) a LISP condition, which indicates when that comment is to be included in the critique, and (2) a pointer to the prose comment itself. Alternatively, as illustrated in the second comment frame above, a simple prose fragment may be stored in the frame. Also, as illus-trated in the third comment frame, an entire expressive frame can be activated.

In addition, the comment frame can contain any other descriptive information about the comment. This information is used to help coordinate the processing of the expressive frame.

7.7.4. The Function That Coordinates Processing of the Expressive Frame

The third element of an expressive frame is the function that will coordinate the processing of the comment frames. The user may specify his own function or may use the builtin function GENCOMMENTS as illustrated in the example above. GENCOMMENTS performs the following:

1. It evaluates the conditions of each comment frame to see which will be included in the critique.
2. It sorts these according to an associated sequence indicator.
3. It activates the prose generator on each comment in turn (in the sorted sequence), thereby creating the critique.

The system designer may also define his own function in place of GENCOM-MENTS, giving him more flexibility in structuring the critique at the paragraph level. The next section gives an example of how this is done.

7.8. Using PROSENET at the Paragraph Level

The previous section described how a critique can be produced by generating entire sentences, one after another. This approach is satisfactory for certain cri-tiques but does not allow flexible paragraph-level structuring of the prose. This section shows how a paragraph-level PROSENET network can be used to coor-dinate prose generation. A paragraph-level PROSENET network may function in two general ways:

1. It can insert appropriate connective words and phrases around whole sen-tences.
2. It can piece a paragraph together around sentence fragments, as shown in the example below.

The following expressive frames is used by the VQ-ATTENDING system (P. L. Miller 1985a) to discuss active oxygenation treatment goals for a patient receiv-ing mechanical respiratory support.

```
(DEFFRAME 'OX_GOALS '(

(IF (SAME ACHIEVE_PO2_GOAL ACTIVE) GOAL ACHIEVE_PO2_GOAL
        COMMENT (to achieve an adequate paO₂))

(IF (SAME MAINT_PO2_GOAL ACTIVE) GOAL MAINT_PO2_GOAL
        COMMENT (to maintain an adequate paO₂))
```

```
(IF (SAME AVOID_O2_TOX_GOAL ACTIVE) GOAL AVOID_O2_TOX_GOAL
        COMMENT (to avoid the risk of oxygen toxicity))

(IF (SAME REDUCE_O2_TOX_GOAL ACTIVE)
        GOAL REDUCE_O2_TOX_ GOAL
        COMMENT (to reduce the risk of oxygen toxicity))

(IF (SAME AVOID_PEEP_GOAL ACTIVE) GOAL AVOID_PEEP_GOAL
        COMMENT (to avoid the risks associated with high PEEP))

(IF (SAME REDUCE_EXC_PEEP_GOAL ACTIVE)
        GOAL REDUCE_EXC_PEEP _GOAL
        COMMENT (to reduce the risks associated with high PEEP))

(IF (SAME REDUCE_O2_SUPPORT GOAL ACTIVE)
GOAL REDUCE_O2_ SUPPORT_GOAL
        COMMENT (to reduce the level of oxygenation support))

(IF (SAME MAINT_FIO2_GOAL ACTIVE) GOAL MAINT_FIO2_GOAL
        COMMENT (to maintain FiO₂ at maintenance levels *openparen "0.2 to
                0.4" *closeparen))

(IF (SAME MAINT_PEEP_GOAL ACTIVE) GOAL MAINT_PEEP_GOAL
        COMMENT (to maintain PEEP at maintenance levels *openparen "0 to
                5" *closeparen))

)

'OXDISS)
```

This expressive frame has nine comment frames. Notice that each contains a prose fragment rather than a pointer to a PROSENET network. This expressive frame also has a *user-defined* function (OXDISS) to coordinate its processing. OXDISS is defined below:

```
(DE OXDISS (COMMLIST)
        (PROG (GOALS HIGH)
        (SETQ GOALS (SELCOMMENTS COMMLIST))
        (SETQ HIGH (FINDHIGH GOALS))
        (SETQ GOALS (REMOVE HIGH GOALS))
        (GENPROSE '$OXGOALS)
        ))

(DE FINDHIGH (COMMLIST)
        (COND ((NULL COMMLIST) NIL)
                ((SAME (GET1 (CAR COMMLIST) 'GOAL) PRIORITY HIGH)
                (CAR COMMLIST))
                (T (FINDHIGH (CDR COMMLIST)))))

(DE PRINTGOALS (L)
        (PROG( )
```

```
A (COND ((NULL (CDR L))(RETURN (PRINTGOAL (CAR L))))
        ((NULL (CDDR L)) (PRINTGOAL (CAR L)) (RUNOFF
        '(*comma and)))
        (T (PRINTGOAL (CAR L)) (RUNOFF '*comma)))
  (COND ((SETQ L (CDR L)) (GO A))) ))

(DE PRINTGOAL (COMM) (GENCOMMENT COMM))
```

OXDISS calls the builtin function SELCOMMENTS (see Appendix III), which evaluates the IF conditions of each comment frame (each of which is a description of one possible oxygenation treatment goal) and creates a sublist (GOALS) of those treatment goals to be discussed in the critique. This list is then inspected by FINDHIGH to see if one of the goals has "high priority." If so, it is stored in HIGH and deleted from GOALS. Finally the function GENPROSE (see Appendix III) is called to activate the paragraph-level PROSENET network "$OX_GOALS," shown below.

```
(DEFPROSE '(

($OXGOALS (*para $OX1O T))

($OX1O ((*OPTION ((in regard to) $OX2O T)
               ((looking first at) $OX2O T)
               (  T T))

($OX2O ((oxygenation *comma) $OX3O T))

($OX3O ((the primary goal in this patient *apost_s mangement is)
               $OX4O HIGH)
        ((there are several goals for this patient *apost_s management
               *period) $OX2OO (CDDR GOALS))
        ((there are two main goals for this patient *apost_s mangement
               *period) $OX2OO (CDR GOALS))
        ((there is one main goal for this patient *apost_s mangement *colon)
               $OX6O T)
        )

($OX4O  ((*function (PRINTGOAL HIGH)) $OX5O T))

($OX5O  ((*period secondary goals are) $OX6O (CDR GOALS))
        ((*period a secondary goal is) $OX6O GOALS)
        (*pop T T))

($OX6O  ((*function (PRINTGOALS GOALS)) $OX7O T))
($OX7O  (*period $OX24O T))

($OX2OO ((*sequence SEQ1 ((one goal is) $OX21O T)
                        ((a second goal is) $OX21O T)
                        ((a third goal is) $OX21O T)
                        ((another goal is) $OX21O T)
                        )  T T))
```

```
($OX210 ((*function (PRINTGOAL (CAR GOALS))) $OX220 T))

($OX220 (*period $OX230 T))

($OX230 (*jump $OX200 (SETQ GOALS (CDR GOALS)))
        ((*endsequence SEQ1) $POPTT T))

($OX240 ((the urgency of the primary goal *comma however *comma may
         override the secondary considerations *period) $POPTT HIGH)
        (*pop T T))

($POPTT (*pop T T))

))
```

This paragraph-level PROSENET network uses its action routines to test the comment frames stored in GOALS and HIGH, which thereby guide the prose generation. A different paragraph structure is created depending on whether a high-priority goal is found, and depending on how many goals are to be described.

The following LISP statements illustrate how this prose-generating activity is initiated:

```
(DEFFACT '(ACHIEVE_PO2_GOAL ACTIVE))
(DEFFACT '(ACHIEVE_PO2_GOAL PRIORITY HIGH))
(DEFFACT '(AVOID_O2_TOX GOAL ACTIVE))
(DEFFACT '(AVOID_PEEP_GOAL ACTIVE))
(GENFRAME 'OX_GOALS)
```

This sequence results in the following critique:

> In regard to oxygenation, the primary goal in this patient's management is to achieve an adequate paO$_2$. Secondary goals are to avoid the risk of oxygen toxicity, and to avoid the risks associated with high PEEP. The urgency of the primary goal, however, may override the secondary considerations.

A paragraph-level PROSENET network gives the system designer great flexibility in coordinating the creation of a critique. Expressive frames allow each comment to be inspected and tested prior to prose output. The ability to couch these comments within a flexible prose structure at the paragraph level greatly enhances PROSENET's expressive power.

7.9. Top-Level Routines to Coordinate the Critiquing Analysis

The three components of E-ATTENDING described in Sections 7.5, 7.6, and 7.7 allow a system builder to implement the heart of a critiquing system. Depending on the domain, however, a significant amount of additional system implementation may be required. Additional routines to complete a critiquing system fall into two general categories.

1. Data-gathering routines. As discussed in Section 7.3, the routines which initially gather information from the physician (describing the patient and the proposed approach), must be written by the system builder.
2. Top-level routines. The system builder must also write the top-level routines that coordinate the operation of the critiquing system. These routines:
   a. Introduce the system to the user
   b. Call the data-gathering routines
   c. Activate the production rule interpreter
   d. Activate the various expressive frames appropriately so that the components of the physician's plan are critiqued in sequence

In addition to these two types of routines which must be added to a critiquing system built using E-ATTENDING, the designer may choose to augment E-ATTENDING's capabilities in various ways, as discussed below in Section 7.10.2.

7.10. E-ATTENDING: Current Status and Possible Extensions

As mentioned previously, the current version of E-ATTENDING has been deliberately left simple from a number of standpoints. This design decision raises two questions. (1) To what critiquing domains can the current version of E-ATTENDING be applied unchanged? (2) What refinements and augmentations might be needed to implement the critiquing approach in more complex domains? This section discusses each of these questions in turn.

7.10.1. The Current E-ATTENDING: Appropriate Domains

Whether a domain is appropriate for the current E-ATTENDING depends on certain characteristics of the domain. As discussed more fully in Chapter 9, in some domains each critiquing comment is able to react "locally" to a small number of input variables. For example, in the workup of a pheochromocytoma, if the physician suggests a glucagon stimulation test for a hypertensive patient, a critiquing system can comment (1) that this is inappropriate, and (2) that a clonidine suppression test should be used instead. Both of these comments react to a small subset of input data describing the patient and the plan: (1) that the patient is hypertensive, and (2) that a glucagon stimulation test has been ordered. No more comprehensive analysis is needed to produce these comments. Such a domain fits naturally into the current E-ATTENDING design.

In other medical domains, however, the physician may be faced with a wider range of alternatives, each with many potential risks and benefits in the presence of different medical problems. Here, the system may have to examine all the choices and assess the risks and benefits of each in that patient. A large number of variables may need to be tested to determine whether one choice is preferable to another. It would be awkward to test all these variables in E-ATTENDING's expressive frames.

An example of this type of domain was discovered in implementing the ATTENDING system, which critiques anesthetic management. In this domain, the system's comments cannot be designed to react to isolated aspects of the physician's plan. ATTENDING must look comprehensively at all the risks involved in several choices to select those which minimize risk in the patient described.

7.10.2. The Current E-ATTENDING

The current E-ATTENDING is designed for domains where each critiquing comment reacts to a small number of input variables: domains where the number of alternative approaches is constrained, where the number of risks and benefits are limited, and where the system can therefore be programmed a priori with comments to make in the various circumstances that may arise.

This restriction is a major one. Nevertheless a number of interesting medical domains can be handled by such a critiquing system. For instance:

1. Domains of patient workup. There are a number of domains of laboratory and radiologic workup where the number of alternative approaches is very constrained. Indeed, patient workup may be a fertile area for implementing critiquing systems, since many workup problems are very focused. PHEO-ATTENDING is an example of such a system.
2. Domains of medical management. Although some areas of management require a comprehensive analysis of alternatives, other areas do not. For example, HT-ATTENDING and VQ-ATTENDING are currently implemented using E-ATTENDING.
3. Domains of differential diagnosis. As described in Section 1.5, we are currently building ICON, a system to critique differential diagnosis. Although this project is in its early stages, the current E-ATTENDING seems well suited to this problem.

Thus, there are a number of interesting medical domains where a critiquing advisor can be implemented using the current E-ATTENDING.

7.10.3. Possible Extensions to E-ATTENDING

There are several augmentations that might help E-ATTENDING deal with a broader class of domains, including:

1. Hierarchical structuring. Many treatment choices have a hierarchical structure. For instance, in the domain of essential hypertension, the choice of a diuretic involves four levels of decision: (1) *diuretic*, (2) *class* of diuretic, (3) *agent*, and (4) *dose*. A system may critique the use of a diuretic at any level of this hierarchy.

    In the ATTENDING system, the augmented transition network formalism is used to give the system a hierarchical decision structure. It is not clear, however, that the full power of this ATN formalism was needed. In the HT-

. ATTENDING system, the hierarchical structure is captured in an ad hoc fashion. (Expressive frames generate the higher level comments before the lower level comments.) No formal structuring was needed.

As a result, no formal hierarchical formalism has been included in E-ATTENDING. In certain domains, however, an organized hierarchical structuring may prove useful.

2. Risk analysis. As discussed in Section 7.10.1, E-ATTENDING is designed for domains where each critiquing comment need only react to a limited set of input variables. In other domains, however, a system may have to perform a more comprehensive analysis, for instance, of the relevant risks and benefits.

It might therefore be useful to augment E-ATTENDING with a builtin capability to perform such an analysis of risk. In fact, ATTENDING already implements a heuristic approach to risk analysis. Incorporating a similar risk-analysis capability into E-ATTENDING might help it handle more complex domains. This has not been done because it is not yet clear how best to do this in a general way.

7.11. E-ATTENDING: Summary

There are a number of medically interesting domains in which critiquing need not be computationally complex, once the system design issues are understood. The E-ATTENDING system is designed to help researchers build critiquing systems in these domains.

In addition, there are more complex problems posed by other domains which may require more sophisticated critiquing capabilities. The current version of E-ATTENDING provides a starting point, from which researchers may explore how the concepts of critiquing can be expanded and refined.

Chapter 8

ESSENTIAL-ATTENDING's Knowledge Exerciser Program

Implementing an expert computer system in a real-world domain is a formidable task, even if a restricted domain is chosen. A major problem is "validating" the system's knowledge, assuring that it is accurate, consistent, and complete.

One problem in any validation of a medical expert system is that there is seldom a "gold standard" for comparison. Expert physicians themselves frequently approach a given problem differently. A survey of these issues and of the evaluation of AIM systems is given in (P. L. Miller 1985b).

The most obvious approach to knowledge validation is to run test cases through an expert system and assess the results. A different approach, however, is the recent development of tools that let the system itself help in validating its own knowledge. To allow this, a system's knowledge must be stored in a flexible, manipulable format, and the system must be given, in some sense, knowledge about its own knowledge.

Three examples of experimental knowledge validation tools in medical domains are:

1. The TEIRESIAS system (Davis and Lenat 1982), a pioneering exploration into how an expert system might interactively assist in inspecting and updating its own knowledge.
2. ONCOCIN's Rule Checking Program (Suwa et al. 1984), designed to help check a set of production rules for consistency.
3. The SEEK program (Politakis and Weiss 1984), designed to test the performance of diagnostic production rules against a group of cases with known diagnoses. From this analysis, SEEK proposes "experiments": modifications in its knowledge that might enhance its performance.

The motivating factor behind the development of knowledge validation tools is the complexity of an expert system's knowledge. A system designer rapidly loses track of the exact contents of his knowledge base and of how the various pieces of knowledge will interact in different circumstances. As a result, there is a need for tools for debugging such knowledge, just as there are currently tools for debugging more conventional programs. The more sophisticated these debugging aids are, the more help they will be.

This chapter describes a simple augmentation of ESSENTIAL-ATTENDING that affords the domain expert considerable flexibility in inspecting, manipulating, and validating the knowledge of a critiquing system.

8.1. The ESSENTIAL-ATTENDING Knowledge Exerciser

A "knowledge exerciser" program has been developed to assist in validating an ESSENTIAL-ATTENDING knowledge base. The program is domain independent since it operates (with certain restrictions) on any knowledge base built using ESSENTIAL-ATTENDING. The program is also quite simple, comprising approximately two pages of LISP code. The entire knowledge exerciser program is shown in Appendix VI.

The knowledge exerciser program operates as follows:

1. It is passed as its initial input a list of ESSENTIAL-ATTENDING expressive frames.
2. It displays this list as a menu to the domain expert who is asked to select one of the frames.
3. The program then examines that frame and compiles a list of all the conditions that control the prose which the expressive frame outputs. This is done by collecting the IF-clauses from each comment frame, and by exploring all the associated PROSENET networks. Any compound conditions (involving ANDs or ORs) are broken down into their component conditions.
4. A menu of all these conditions is displayed to the domain expert, who may specify any combination of conditions as a "test case" to exercise the expressive frame.
5. The specified conditions are set, and the expressive frame is activated. This typically results in one or more prose paragraphs being generated (an excerpt of a critique). The domain expert can then see exactly how the expressive frame reacts to that particular combination of conditions.
6. The domain expert may then quit, may return to step 4 to test the same expressive frame with a different set of conditions, or may return to step 2 to select a different frame.

Using the knowledge exerciser program, the domain expert can quite quickly exhaust all meaningful combinations of conditions associated with an expressive frame. In this way, each expressive frame can be exhaustively exercised independently of the rest of the system. This process is much easier, quicker, and more comprehensive than running a group of test cases through the entire system.

8.2. An Example: Exercising HT-ATTENDING's Knowledge Base

This section shows how the knowledge exerciser program operates on HT-ATTENDING's knowledge of hypertension management. First, the program introduces itself and displays a menu of HT-ATTENDING's expressive frames. These frames correspond to the different drugs used for treating essential hypertension. (In this example, responses typed by the user are underlined. Explanatory comments are enclosed in square brackets.)

This is a fairly primitive augmentation of HT-ATTENDING, which allows a domain expert to test the various components of the system's knowledge base to see how the system reacts in different circumstances.

*** which of the following would you like to examine?
1. monotherapy
2. step-overview
3. one-agent
4. thiazide
5. loop
6. kcl
7. no-kcl
8. beta-blocker
9. alphamethyldopa
10. clonidine
11. guanabenz
12. rauwolfia-alk
13. captropril
14. prazosin
15. hydralazine
16. calcium blocker
17. guanadrel
18. guanethidine
19. minoxidil

*** type one (0 to quit) ***8

[In this example, the domain expert selects the beta blocker expressive frame. The program therefore displays a list of all the conditions that affect the system's critique of the use of beta blockers. In preparing this list, the program "pretties" the conditions by converting them to lower case, stripping off parentheses, etc.

In the four examples that follow, the domain expert selects different sets of conditions to help assure that the system is indeed reacting correctly and coherently.]

*** please select a set of conditions:
1. peripheal vascular disease

　　　2. diabetes mellitus
　　　3. asthma or COPD
　　　4. CHF
　　　5. labetolol proposed
　　　6. cardiac conduction delay
　　　7. renal insufficiency
　　　8. pindolol proposed
　　　9. beta-blocker twoagents
　　　10. coronary artery disease
　　　11. hypothyroid
　　　12. depression
*** type one or more (0 to quit) ***3

　　There are now seven beta blockers available in the United States. In a patient with asthma, we would recommend avoiding beta blockers entirely in favor of alternatives as discussed above. Although it is clear that patients with asthma should not generally be given beta blockers, should it be necessary to do so, cardioselective agents (metoprolol and atenolol) may be better tolerated and asthmatic attacks precipitated by beta blockade may be more easily reversed.

*** please select a set of conditions:
　　　1. peripheal vascular disease
　　　2. diabetes mellitus
　　　3. asthma or COPD
　　　4. CHF
　　　5. labetolol proposed
　　　6. cardiac conduction delay
　　　7. renal insufficiency
　　　8. pindolol proposed
　　　9. beta-blocker twoagents
　　　10. coronary artery disease
　　　11. hypothyroid
　　　12. depression
*** type one or more (0 to quit) ***4 12

　　There are now seven beta blockers available in the United States. Clearly, in a patient with decompensated congestive heart failure, all beta blockers should be avoided in favor of alternatives as discussed above. On the other hand, in certain patients with well-compensated congestive heart failure, a beta blocker may be useful. Here, pindolol, which has intrinsic sympathomimetic activity (ISA), may well be the beta blocker of choice. Since depression has occasionally been seen with beta blocker therapy, especially with propranolol, the water-soluble agents nadolol and atenolol, which do not penetrate into the CNS in high concentrations, may be preferred for a patient with a history of depression.

*** please select a set of conditions:
　　　1. peripheal vascular disease
　　　2. diabetes mellitus

   3. asthma or COPD
   4. CHF
   5. labetolol proposed
   6. cardiac conduction delay
   7. renal insufficiency
   8. pindolol proposed
   9. beta-blocker twoagents
   10. coronary artery disease
   11. hypothyroid
   12. depression
*** type one or more (0 to quit) ***7 11 5

There are now seven beta blockers available in the United States. Nadolol and atenolol have the advantage that they give a more predictable blood level and have a longer duration of action than other beta blockers. Both of these drugs can be taken once a day. The dose of both agents, however, must be reduced (perhaps halved) for a patient with renal insufficiency. Also, in a hypothyroid patient, one must be especially cautious when using beta blockers since these can interfere with the action of thyroid hormone and the patient's thyroid replacement requirements may change.

Labetolol is the first drug available in the United States with both alpha and beta blocking activity. Although it has distinct advantages in treating both pheochromo-cytoma and accelerated hypertension, labetolol does not seem to be any more effective than pure beta blockers for most situations.

*** please select a set of conditions:
   1. peripheal vascular disease
   2. diabetes mellitus
   3. asthma or COPD
   4. CHF
   5. labetolol proposed
   6. cardiac conduction delay
   7. renal insufficiency
   8. pindolol proposed
   9. beta-blocker twoagents
   10. coronary artery disease
   11. hypothyroid
   12. depression
*** type one or more (0 to quit) ***3 4 1

There are now seven beta blockers available in the United States. In a patient with asthma, we would recommend avoiding beta blockers entirely in favor of alterna-tives as discussed above. Although it is clear that patients with asthma should not generally be given beta blockers, should it be necessary to do so, cardioselective agents may be better tolerated and asthmatic attacks precipitated by beta blockade may be more easily reversed. Also, clearly, in a patient with decompensated con-gestive heart failure, all beta blockers should be avoided in favor of alternatives as discussed above. On the other hand, in certain patients with well-compensated congestive heart failure, a beta blocker may be useful. A patient with peripheral vascular disease might well benefit from a cardioselective agent since these cause less peripheral vasoconstriction.

8.3. The Current Knowledge Exerciser Program: Strengths and Weaknesses

The current knowledge exerciser program is conceptually straightforward but is nevertheless a powerful tool. This power is a reflection not of the program itself (which is quite simple) but of the flexibility gained by expressing a system's knowledge using the domain-independent ESSENTIAL-ATTENDING format. The current knowledge exerciser program does have several significant limitations.

1. One limitation is that, when testing the knowledge base, the domain expert is testing only a portion of the whole expert system. The interface of the rest of the system to the expressive frames must, of course, be debugged as well.
2. When presenting its menu of conditions, the knowledge exerciser program knows nothing about any interactions between the conditions that it lists. Some conditions may be mutually exclusive. Some may be mutually dependent in different ways. As a result, a random selection of these conditions may not make sense. If presented with a nonsense scenario (for instance, that two conflicting conditions hold, a situation that should never be input to the frame by the system itself), the knowledge exerciser program may well produce a nonsense critique. The domain expert must understand that the knowledge exerciser has no knowledge of any interactions between the conditions listed and will not correct any erroneous combinations he selects. The combinations chosen must make sense in the context of the domain as a whole.
3. A third restriction concerns the use of global variables. If any of the conditions test global variables, or if any prose-generating function tests global variables, the knowledge exerciser program itself will not know what these variables should contain. As a result, any global variables must be initialized by the system designer before the knowledge exerciser program is activated.

ESSENTIAL-ATTENDING's knowledge exerciser program is quite simple and has significant limitations. It could be made more sophisticated in several ways. As long as its limitations are recognized, however, the program can nevertheless provide a system designer with a very useful tool.

Chapter 9

Lessons Learned: Design Parameters for a Critiquing System

The previous chapters have outlined our exploration of the critiquing approach in four domains of quite different character. They also discussed how different domains have helped to highlight different facets of critiquing, and how certain common features have been captured in the domain-independent ESSENTIAL-ATTENDING system. This chapter steps back and tries to abstract some of the more general lessons learned.

The chapter discusses certain general design parameters that influence the design and implementation of a critiquing system. In particular, it suggests that the computational process of critiquing can assume a different character depending on the nature of the domain.

The previous chapters have dealt with these issues only in a piecemeal fashion. We have seen, for instance, that in the domain of anesthesiology the ATTENDING system is designed explicitly to assess and compare risks, while in the other systems this is not done. Similarly, in the domain of ventilator management it is useful to give the system explicit knowledge of treatment goals and to let this knowledge drive the critiquing process, whereas in the other systems this is apparently not needed.

Thus, different domains seem to make different demands on a critiquing system. Do these different features merely reflect idiosyncracies of particular domains, or can one make more general statements as to the nature of decision making across different domains?

9.1. Two Dimensions Along Which Domains Vary

We believe that these various differences reflect fundamental differences between the domains, and that different domains place different demands on a critiquing system's design. This chapter suggests two general dimensions along which domains may vary.

1. Local vs. global criteria. In some domains, decisions are "factorable" into independent subdecisions. In these domains, a system may critique a given choice by only looking at a small number of locally related parameters. In other domains a much more global, comprehensive analysis of the entire plan may be required before any single subdecision can be discussed appropriately.
2. Depth of biomedical knowledge. In many domains, patient management is performed by following "established approaches" that have evolved over years of clinical practice by many physicians. In these domains, a critiquing system does not have to reason from fundamental biomedical principles. Instead, the system need only help the physician choose from a predetermined set of alternatives. In other domains, however, a system may have to fall back to more fundamental biomedical knowledge, i.e., "deeper" knowledge about the domain.

The remainder of this chapter discusses both of these dimensions in detail. In so doing, the chapter has two goals. One goal is to develop an understanding of general design parameters that influence critiquing, as outlined above. A more fundamental goal, however, is to explore the underlying process of medical management itself.

Thus, the chapter is more than an attempt to abstract certain general characteristics of critiquing. It also explores why a general model of the process of medical management has proved difficult to develop and outlines two dimensions of complexity which such a model must include.

9.2. One Dimension: Critiquing Based on Local vs. Global Criteria

The first dimension along which a critiquing system may vary involves the degree to which local vs. global criteria are required to critique a particular decision. Along this dimension, three types of critiquing can be identified: (1) critiquing by reacting to local criteria, (2) critiquing by local risk analysis, and (3) critiquing by global plan analysis.

9.2.1. Critiquing by Reacting to Local Criteria

The most straightforward form of critiquing is critiquing by reacting to local criteria. As illustrated in Figure 9.1, such a system operates as follows:

1. Data are gathered from the physician describing a patient and a proposed management plan.
2. Further inferences are drawn from these data.
3. Via a fairly straightforward mapping, the material to be included in the critique is selected on the data and inferences.
4. This material is assembled into the prose critique.

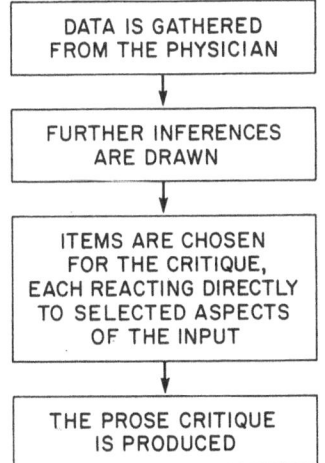

Figure 9.1. Critiquing by reacting to local criteria: Overview.

The most important feature here is that each item included in the critique is a direct reaction to a limited subset of the data gathered. This contrasts with domains where a system must perform a more comprehensive analysis to determine which comments should be included in its critique.

Critiquing by reacting is feasible only in domains where the range of management alternatives is small and the number of risks and benefits associated with each choice is also small. For instance, each decision might involve at most two or three alternatives, each with one or two potential risks. In such a domain, if the physician recommends the "wrong" choice, the system need not perform a comprehensive analysis of many choices to make appropriate comments and propose an alternative.

PHEO-ATTENDING is an example of a system implemented in such a domain. For instance, if (1) an initial plasma catecholamine test yields an equivocal result, and (2) a physician suggests a glucagon stimulation test for (3) a hypertensive patient, PHEO-ATTENDING *reacts* by producing comments as follows:

1. It agrees with the choice of an initial plasma catecholamine test.
2. It agrees that further workup is appropriate after the plasma catecholamine result is equivocal.
3. It disagrees with the physician's plan to order a glucagon stimulation test in a hypertensive patient.
4. It recommends a clonidine suppression test instead.

(In addition, the system indicates its rationale for these recommendations.) Notice that each reaction is a direct response to a specific subset of information about the patient and the workup. PHEO-ATTENDING can critique in this reacting fashion because the number of alternatives, and risks, at each point is very limited.

Since critiquing by reacting is such a straightforward form of critiquing, it is the easiest form to implement. The approach is only feasible in a subset of possible domains. It is nevertheless anticipated that the approach will prove appropriate for many medically interesting domains. ESSENTIAL-ATTENDING is designed to implement this class of critiquing systems.

9.2.2. Critiquing by Local Risk Analysis

In other domains, however, there is a fairly broad range of possible choices at different stages of a patient's management. These alternatives usually exist because no choice is ideal. Each has potential risks and benefits in the presence of different medical problems. To critique such a patient's management, as shown in Figure 9.2, a system operates as follows:

1. It gathers information from the physician.
2. It makes further inferences based on those data.
3. For each decision and subdecision, it explores all the alternative approaches to the patient's management and determines the risks and benefits of each.
4. It weighs these various risks and benefits against one another. In the process, certain approaches may be clearly superior to the proposed approach. Others may be clearly inferior. Still other approaches may be roughly comparable but involve a different set of risks and benefits.
5. The prose critique is then assembled. The critique mentions (1) any risks and benefits of the proposed approach, (2) any superior or comparable approaches along with the risks and benefits of each.

9.2.2.1. How Critiquing by Risk Analysis Differs From Critiquing by Reacting. The central component of the process outlined above is a comprehensive analysis of the risks involved in each decision. This risk analysis is performed by the

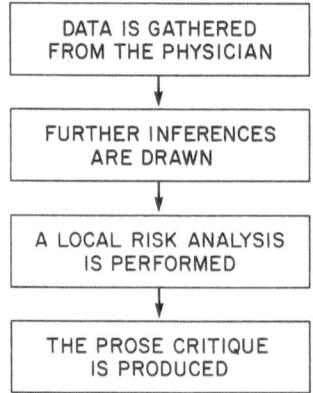

Figure 9.2. Critiquing by local risk analysis: Overview.

system itself to tailor its recommendations to the particular patient and management plan.

Had there been a restricted set of choices (e.g., two choices), then the "critiquing by reacting" approach could be used. The various conditions that make each approach preferable could be listed explicitly. The system designer could program the system to "react" appropriately to each set of conditions. He himself would therefore perform a comprehensive risk analysis when building the system and compile the results of his analysis into rules to guide the "reacting" process.

With a larger number of choices (and a larger number of possible risks and benefits), however, any attempt to precompile all the combinations of conditions into a "reacting" system would become unwieldy. Instead, a general mechanism to perform the risk analysis dynamically is required. To accomplish this task, a general heuristic approach to risk analysis was developed in the ATTENDING system (P. L. Miller, 1983a, 1983b, 1984).

[A similar problem has been encountered previously in two noncritiquing systems. In MYCIN's treatment selection algorithm (Clancey 1984) and in an occular herpes treatment advisor (Kastner et al. 1982), it was found that rule-driven inferencing (which is analogous to critiquing by reacting) became unwieldy in the face of several treatment options, each with several indications and contraindications. A more comprehensive ability to compare the different approaches was required.]

9.2.2.2. Critiquing by Risk Analysis: Appropriate Domains. As described above, critiquing by risk analysis is required in domains that have a range of management options, each with risks and benefits in the presence of different medical problems. ATTENDING operates in such a domain. A number of choices modeled in ATTENDING allow five or more alternatives, and many alternatives have ten or more potential risks in different patients.

In producing its critique, ATTENDING does not try to preprogram all these conditions to let it "react" to every possible combination of conditions. Instead, as described in Chapter 3, it performs a comprehensive risk analysis of all possible approaches to each aspect of the plan.

9.2.3. Critiquing by Global Plan Analysis

Even though the ATTENDING system performs a dynamic risk analysis in constructing its critique, it still does not perform a totally global analysis of the plan to critique each individual choice. Instead it assumes that each subdecision can be analyzed independently and that subdecisions made in one part of the plan do not affect subdecisions made elsewhere. In other words, it assumes that the plan as a whole is "factorable" into independent subdecisions.

On the other hand, in certain domains, a global analysis of the whole plan may be required before a particular subdecision can be discussed. None of the domains discussed in this book had that character. As a result, the problem of constructing a critiquing system to operate in such a domain remains for the future.

9.3. The Second Dimension: Depth of Biomedical Knowledge

The second dimension along which a critiquing system may vary concerns the depth of biomedical knowledge which is required. Here again, three levels of knowledge can be identified (see Figure 9.3): (1) established approaches, (2) treatment goals, and (3) fundamental biomedical principles.

This section describes how different domains require a critiquing system to operate at different levels of medical knowledge. To help make the discussion concrete, examples are given from the domain of essential hypertension.

9.3.1. Critiquing Based on Established Approaches

In many areas of medicine, a physician's treatment is not derived by de novo reasoning from underlying biomedical principles, or even treatment goals. Instead, it is based on formal or informal "established approaches" that outline the various possible treatment steps. Examples of such established approaches are:

1. In oncology, the domain of ONCOCIN (Shortliffe et al. 1981), the administration of chemotherapy (anticancer drugs) is dictated by elaborate official protocols that a physician is asked to follow. These protocols may be over 50 pages in length and attempt to specify in detail exactly how chemotherapeutic drugs are to be given. Oncology is almost unique in medicine in the use of such rigid protocols. Virtually all other domains allow the physician much more freedom.

2. In the management of essential hypertension, a concept of "stepped care" has evolved. Certain drugs are designated "step one" drugs, to be used for initial therapy. Only if these drugs are unsuccessful is a "step two" drug added, and so on to "step three" and finally "step four," the final stage. There is fair, but far from total, agreement as to which drugs belong in each step category. Generally, drugs with lesser side effects are used early, and more potent drugs

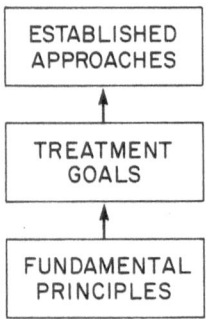

Figure 9.3. Three levels of knowledge that underlie medical management.

with more significant side effects are reserved for later use if needed. There is, however, considerable leeway in the application of the stepped-care approach, and great practice variation. Thus the stepped-care approach is a loosely structured guideline.

These established approaches might be perceived as a "compiled" form of deeper knowledge. Certainly the agents used can be justified in terms of underlying principles and treatment goals. On the other hand, the established approaches have an independent validity of their own, since they reflect years of clinical experience and evaluation.

Another reason these established approaches have independent validity is that the drugs and treatments used are themselves imperfect. They all have potential side effects, which may be totally independent of their therapeutic action. In many domains, established approaches are designed as much to minimize these extraneous side effects as to accomplish the desired therapeutic goals. Indeed, when following an "established approach," a physician may be thinking more about minimizing idiosyncratic side effects than about underlying principles and treatment goals.

As a result, a physician does not reason from fundamental biomedical principles every time he manages a patient. In fact, it would be inappropriate to do so, since it would ignore years of accumulated experience as to (1) appropriate therapeutic goals, and (2) appropriate tactics for achieving those goals.

If a physician is asked why he has taken an action, he may explain that action by citing underlying principles and treatment goals. He may actually select his action, however, from a predetermined set of alternatives to minimize possible side effects. The construction of the set of alternatives has occurred incrementally over years of practice by many clinicians.

Three of the critiquing systems described in this book operate at the level of established approaches: ATTENDING, HT-ATTENDING, and PHEO-ATTENDING.

1. In ATTENDING's domain of anesthesiology, the various agents used for the different stages of anesthetic management are well established. For example, there is a clear-cut set of agents used for induction, intubation, and maintenance of anesthesia. The main problem is making sure that an appropriate choice is made for a particular patient.

2. Similarly, as discussed above, an established stepped-care approach has evolved in HT-ATTENDING's domain of essential hypertension. This approach is structured to minimize side effects, and to allow a patient's blood pressure to be controlled by as simple and benign a regimen as possible. Although the stepped-care approach is far from rigid, it nevertheless imposes considerable structure on hypertension management decisions.

3. Domains of workup, such as PHEO-ATTENDING's, also provide very structured established approaches. This is because very elaborate, specific tests are available.

Thus, most of the critiquing described in this book occurs at the level of *established approaches*. Indeed, such structured problems may well be the most promising initial domains for expert computer advice.

9.3.2. Critiquing Based on Treatment Goals

An intermediate level of knowledge that may be used by a critiquing system involves treatment goals. For instance, in the management of essential hypertension, these goals include:

1. Reducing systemic vascular resistance
2. Suppressing myocardial contractility
3. Suppressing central sympathetic nervous system stimulation
4. Maximizing patient compliance.

In a sense, these therapeutic goals represent a distillation of the knowledge of fundamental principles, since the goals can presumably be derived from that knowledge. Thus, here again, these goals might be thought of as a compiled form of the deeper knowledge. Again, however, these goals have an independent validity in and of themselves, since they are derived from years of clinical experience treating patients and are supported by many clinical studies. If one were to inspect the fundamental principles, and from these hypothesize all plausible treatment goals, some of the proposed goals might have been found to be irrelevant to clinical practice. These would therefore not be included at this level of knowledge.

Thus another form of critiquing is goal-directed critiquing, where the analysis is guided by an assessment of treatment goals. Here, before critiquing the physician's plan, a system first deduces the treatment goals that *it* considers appropriate for the patient's management. It then uses those treatment goals to drive its critiquing analysis.

As illustrated in Figure 9.4, such a system operates as follows:

1. Data are gathered from the physician.
2. Inferences are drawn from these data. Included in these inferences are a set of treatment goals for that patient's management.
3. These inferred treatment goals are discussed explicitly in the prose critique and are also used internally to help select material for the critique of the plan itself.
4. The prose critique of the plan is produced.

VQ-ATTENDING is the only current goal-directed critiquing system. (It operates in a domain where the alternative modalities are limited and, therefore, seen along the other dimension of critiquing, VQ-ATTENDING also critiques by reacting.)

As discussed in Chapter 5, goal-directed critiquing seems to be most useful in domains where management choices are not discrete. Most domains of medicine involve a discrete set of alternatives: e.g., to use drug A, B, or C. In the domain

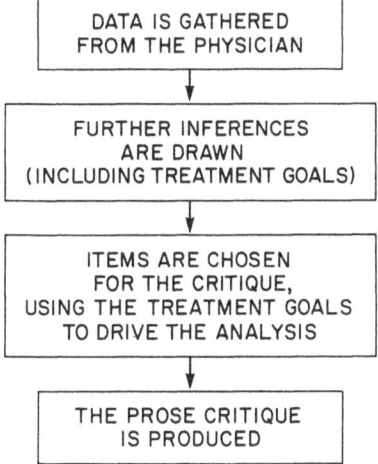

Figure 9.4. Goal-directed critiquing: Overview.

of ventilator management, however, management choices must be selected from a continuum of possible values. Fractional inspired oxygen (FiO_2) can vary from 0.2 to 1.0; positive end-expiratory pressure (PEEP) can vary from 0 to 30 or 40 cm H_2O. Respiratory rate and tidal volume are also chosen from a continuum of values. In the face of this continuum of choice, it proved useful to let VQ-ATTENDING's critiquing analysis be driven by an explicit assessment of treatment goals.

9.3.3. Critiquing Based on Fundamental Biomedical Principles

The most fundamental level of knowledge includes the biomedical principles that underlie the domain: knowledge about physiology, anatomy, pharmacology, and also patient behavior. In the management of essential hypertension, this level includes:

1. Knowledge of the role of renal function in controlling intravascular volume and electrolytes
2. Knowledge of the cardiovascular system and the various pharmacologic and phyisologic parameters that manipulate it
3. Knowledge of the central and peripheral nervous system as it relates to the control of blood pressure.

Most clinical decisions in the management of essential hypertension, however, are not made at this level. Indeed, the main utility of this knowledge may well be for pharmaceutical researchers developing new drugs to treat hypertension.

Nevertheless, there may be times when knowledge at this level is used by the practicing physician. For example, in a very difficult hypertensive patient, a hypertension expert may fall back to deeper levels of reasoning. Similarly, in a hypertensive patient with other medical problems, a physician may at times modify the medical treatment of these other problems, hoping indirectly to control blood pressure. In so doing, he may draw directly on knowledge of fundamental biomedical principles.

The four ATTENDING systems currently implement critiquing at the level of established approaches (ATTENDING, HT-ATTENDING, PHEO-ATTENDING) and therapeutic goals (VQ-ATTENDING). None incorporates an explicit knowledge of fundamental principles.

9.4. How Deeper Knowledge Might Be Used by a Critiquing System

There are several different contexts in which deeper knowledge might be integrated into a critiquing system. This section discusses three ways in which this might be done:

1. Critiquing beyond the scope of an established approach
2. Using the clinical literature as a deep model of the critiquing process
3. Critiquing a physician's choice of treatment goals

9.4.1. Critiquing Beyond the Scope of an Established Approach

Although many areas of medical management have evolved structured established approaches, there are areas of medicine where this has not occurred. (1) A domain may be too complex or too variable for even a loosely structured approach. (2) A patient may exhaust the management options of a standard approach. (3) A patient may have interrelated problems that make established approaches suboptimal.

In any of these situations, an expert system may have to fall back on deeper levels of reasoning, for instance, based on underlying biomedical principles. The nature of such deep analysis, however, is unclear and is an open area for research.

9.4.2. Clinical Literature as the Deep Model of Decision Making

In expert system research, the term "deep knowledge" is usually used to refer to knowledge of fundamental biomedical principles underlying a domain. Several research projects have explored how best to incorporate such knowledge into medical systems (Patil et al. 1981; R. A. Miller et al. 1982). Since medical management is often guided by established approaches, however, there is a different type of deep knowledge that governs much of clinical practice: a critical understanding of the relevant clinical literature.

In clinical practice, treatment decisions are frequently based on accumulated experience reported in the literature, e.g., studies evaluating the relative efficacy of different established approaches to a clinical problem. Several different studies may apply to a given treatment decision. Moreover, these studies may conflict as to their conclusions.

To model this knowledge of clinical literature, a computer would need a variety of information about each clinical study, including:

1. The specific clinical decision to which the study applies
2. The patient population to which it applies
3. Any relevant methodologic weaknesses in the study
4. A critical evaluation of how the implications of one study interacts with others

Such information is tailor made for use by a critiquing system, since a clinical study is often an experimental critique of a particular management choice. A critiquing system might use this "deep knowledge" of the clinical literature in several ways:

1. Given a patient and a treatment decision, it could critique that decision with a reasoned analysis supported by relevant clinical studies.
2. Given a patient, a treatment decision, and a specification of possible relevant articles, it could critique the relevance of the articles to the treatment decision.
3. Finally, since medical knowledge constantly evolves (as new treatments are developed and new studies are published), the system might be designed to help incorporate this new knowledge into its knowledge base, and therefore into its critiquing analysis.

In summary then, a critical knowledge of clinical literature could be particularly useful when a physician is managing a patient using an established approach. It also demonstrates one way in which a form of deeper knowledge could be incorporated into a critiquing system.

9.4.3. Critiquing Treatment Goals

Chapter 5 discussed how a critiquing system (VQ-ATTENDING) can base its analysis on a set of treatment goals that *it* considers relevant to the patient described. A potential problem may occur, however, if the physician is using a different set of treatment goals.

For instance, even physicians may disagree about satisfactory arterial oxygenation in a given patient. One physician might consider a patient's oxygenation to be unsatisfactory and therefore have the goal of achieving adequate arterial oxygenation. Another physician might consider that patient's oxygenation to be satisfactory and therefore have the goal of maintaining an adequate oxygenation. This problem is particularly likely to occur when a patient's oxygenation is borderline.

As a result, an expert system and its physician user may not always agree as to appropriate treatment goals. One solution to this problem is to allow the system to critique the physician's plan at two levels: (1) at the level of management alternatives, as VQ-ATTENDING does now, and also (2) at the level of treatment goals. In other words, the system could critique the physician's treatment goals as well as his tactics to achieve those goals. Such a system might operate as follows:

1. First it might develop a model of the physician's own treatment goals, by deducing them from his plan and/or by asking.
2. It might then compare the physician's treatment goals with its own inferred goals and critique any salient differences.
3. Finally, it might critique the physician's plan from the perspective of both sets of goals.

Designing a system to critique treatment goals could draw upon deeper knowledge of underlying biomedical principles. For instance, to critique the appropriateness of a given arterial oxygenation, the system might discuss:

1. The oxygen content of blood as a function of the hemoglobin–oxygen dissociation curve
2. Factors that influence actual delivery of oxygen to the tissues, including cardiac output
3. Ventilation–perfusion relationships in the lung

In summary, then, designing a system to identify treatment goals explicitly points up a new form of critiquing: the critiquing of those goals themselves. The critiquing of treatment goals could draw upon fundamental biomedical knowledge and would therefore demonstrate a further way in which such knowledge could be incorporated into a critiquing system.

9.5. Summary: Toward a General Model of Medical Management

Over the past 15 years a number of expert systems have been developed applying artificial intelligence techniques in areas of both medical diagnosis and management. From this research, several well-defined models of medical diagnosis have been developed. These include rule-driven inferencing models, as in MYCIN (Shortliffe 1976) and EXPERT (Weiss and Kulikowski 1979), and "hypothesize and test" models as in INTERNIST-I (R. A. Miller et al. 1982).

The process of medical management, however, has not been as easily captured in a general model. Although several projects have dealt with areas of medical management, a general conceptual framework has not emerged. Indeed, a single unifying model of the decision-making process underlying medical management may not be feasible. (On the one hand, one might suggest that optimal management decisions should be constructed from a full consideration of underlying biomedical principles. As discussed above, this may be necessary in some cases. On the other hand, many management decisions involve the more superficial task of choosing from a predetermined set of established alternatives.)

This chapter has therefore outlined a general framework for thinking not only about critiquing, but also about the process of medical management as a whole. In particular, the chapter makes two observations. (1) The conceptual process of medical management may occur along two dimensions, as discussed above. (2) The particular characteristics of a domain may force an expert system to operate at different levels of this spectrum. In fact, this may be why it has been difficult to abstract a general model of the underlying process of medical management. The chapter has used the critiquing approach as a vehicle to explore these issues.

Chapter 10

Future Directions

This final chapter has two parts. The first part summarizes the current status of the critiquing approach, as outlined throughout this book. The main subject of the chapter, however, concerns the future. Where is research into critiquing likely to go from here?

1. What are promising areas for further research?
2. What further sophisticated critiquing capabilities might be explored?
3. What issues must be confronted in the validation and clinical evaluation of a critiquing system?
4. What mechanisms might be most appropriate for disseminating critiquing technology for practice use?

In general, we can only speculate as to the general directions in which these questions will lead.

10.1. Critiquing Research: Current Status

The research described in this book represents an initial exploration of system-design issues involved in the critiquing approach. Four prototype systems have been described in four quite different medical domains. The purpose in implementing these developmental systems has been severalfold:

1. To demonstrate that the critiquing approach is widely applicable in medicine
2. To explore various facets of critiquing highlighted most clearly in particular domains
3. To identify common features of critiquing
4. To characterize the general design parameters of the critiquing approach

10.1.1. The Generality of the Critiquing Approach

ATTENDING, our first critiquing system, was implemented in a medical domain that is quite esoteric (anesthesiology). Most physicians know little, if anything, about anesthetic management. As a result, we felt it was important to explore other areas of medical management, if only to demonstrate that critiquing was not a reflection of some idiosyncracy of anesthetic management.

The domain of essential hypertension was chosen for our next system in part because it is such a central problem of primary care, familiar in concept to virtually all physicians. It is also a very prevalent, important, and topical disease. We felt that a system which critiques hypertension management could achieve high visibility and that most physicians would understand the character of its advice.

Our third domain of medical management (ventilator management) was chosen in part because we felt that three systems in very different domains of medical management would constitute an informal "existence proof" that critiquing was indeed a general technique. Ventilator management was particularly appealing because it seemed to have a much different character from the previous two domains. This indeed did prove to be the case, as evidenced by the fact that the domain highlighted so clearly the role that treatment goals might play in a critiquing system.

We originally felt that these three domains of medical management would provide a sufficient demonstration of the potential generality of critiquing. We soon realized, however, that there was another whole area of medical practice where critiquing could be applied: patient workup. This realization led us to extend critiquing into a domain of workup, (1) to explore the design issues involved, and (2) to demonstrate that, indeed, workup could be a productive area for critiquing research. The fact that many domains of workup are quite constrained and that optimizing workup might offer considerable cost savings make workup particularly appealing. As a result, the PHEO-ATTENDING system was implemented to critique pheochromocytoma workup. The real goal of this system, however, was to explore the critiquing of workup in general and to highlight this potential area for critiquing advice.

During the implementation of these four systems in areas of medical management and workup, we had deliberately avoided differential diagnosis. We felt that differential diagnosis was a completely different process from management and workup, and it did not occur to us that the critiquing approach might apply.

We came to realize, however, that critiquing could very easily be applied to differential diagnosis. Indeed, we now feel that giving a physician detailed feedback to help in differential diagnosis may well turn out to be a most important application of critiquing.

As a result, we are currently exploring the critiquing of radiologic differential diagnosis in the ICON system (P. L. Miller et al. 1985). Here again, the emphasis is on exploring the design issues involved in critiquing diferential diagnosis and demonstrating that this is indeed a potential area for critiquing research.

This brings us to the present. With developmental systems in areas of medical management, workup, and differential diagnosis, we hope that the generality of the critiquing approach is convincing. The next step, of course, is to demonstrate the practical value of the critiquing approach in an operational system.

10.1.2. Exploring Different Facets of Critiquing

As discussed throughout this book, the exploration of different domains helps to highlight different facets of the critiquing process. This, in turn, yields a fuller understanding of the fundamental design issues involved. Particular facets of critiquing highlighted in this fashion include:

ATTENDING. The domain of anesthetic management highlight risk and the process of risk analysis as a central paradigm of medical management and of the critiquing approach. This led to the development of the heuristic approach to risk analysis that ATTENDING uses to help assess risk and evaluate risk tradeoffs.

HT-ATTENDING. An important design issue highlighted by HT-ATTENDING was the perception of an expert critiquing system as an interactive paper that tailors its content to a particular approach to a patient's care. This vision of a system's role helps define how such a system might be designed and the type of knowledge it could contain.

VQ-ATTENDING. The central, unique design feature of VQ-ATTENDING was its emphasis on "goal-directed" critiquing and the explicit separation of strategic knowledge of treatment goals from tactical knowledge of management choices. This system demonstrated how treatment goals could be made a central driving component of a critiquing system's analysis.

PHEO-ATTENDING. Aside from its primary focus on exploring workup as an area for critiquing advice, PHEO-ATTENDING highlighted the problem of conflicting expertise. To our knowledge, conflicting expertise has not previously been addressed in an organized, systematic fashion.

Thus, individual domains have served to highlight particular design issues. As the critiquing approach is explored in further domains, other facets will presumably be identified. It is anticipated that once these design issues are well understood (by implementation in one domain), they may well be applied to other domains in ways not initially obvious in that domain alone.

10.1.3. Identifying Common Features in Different Domains

A further purpose in exploring several domains has been to identify common features and to embody these in the domain-independent, system-building system, ESSENTIAL-ATTENDING. As described in Chapter 7, ESSENTIAL-ATTENDING is now being used to implement several critiquing systems, including VQ-ATTENDING and HT-ATTENDING. A major advantage of such a system is that it greatly simplifies the task of implementing an expert critiquing system in appropriate domains.

A further advantage is that ESSENTIAL-ATTENDING gives a critiquing system an organized underlying structure. As a result, the system's knowledge base is better organized and more easily understood by the system designers. For example, the conversion of HT-ATTENDING from its initial implementation to its present ESSENTIAL-ATTENDING implementation has resulted in an enhanced and more elegant internal design. A further advantage of building a system using ESSENTIAL-ATTENDING is that the design facilitates the development of general tools, such as the knowledge exerciser program.

As emphasized in Chapter 7, the current ESSENTIAL-ATTENDING has been deliberately left simple. As different critiquing capabilities become better understood, ESSENTIAL-ATTENDING could be augmented to include these capabilities in a domain-independent way.

10.1.4. Characterizing Fundamental Design Parameters

The final benefit derived from exploring the critiquing approach in several domains is that it has allowed us to characterize certain fundamental design parameters. It has let us better understand how the nature of a domain influences the scope and complexity of the critiquing process.

As discussed in Chapter 9, two general dimensions are identified which influence a critiquing system's design. One dimension is the degree to which local vs. global information is required to critique a particular decision. The second dimension is the depth of medical knowledge required: knowledge of established approaches, of treatment goals, or of fundamental biomedical principles.

Different critiquing systems may fall at different places along each of these two dimensions, depending on the particular characteristics of decision making in their domains.

10.1.5. Summary: Rationale for a Research Strategy

By exploring several domains, we now have a much better and more comprehensive understanding of the critiquing process than had we narrowed our initial focus. In particular, we have a fuller feeling for: (1) the spectrum of design issues involved, (2) a good domain-independent design, (3) domains that may be productive for practical critiquing advice, and (4) specific domain characteristics that influence critiquing system design. With this fuller perspective, we are better equipped to identify productive domains for further investigation, to implement well-engineered systems in those domains, and to identify productive directions for future research.

10.2. Future Research Areas

10.2.1. Extending Critiquing to Differential Diagnosis

An interesting future research direction is extending the critiquing approach beyond medical management and workup into areas of differential diagnosis. As discussed previously, one project is already underway in radiology. Here, the goal is to give the physician feedback as to whether (and why) his diagnostic conclusions make sense or not.

To critique differential diagnosis, a system would ask the physician to describe the relevant findings he had obtained: symptoms, signs, and preliminary laboratory data. It would also ask for his proposed diagnosis. The system would then critique this as follows:

1. It would give a rough estimate of how strongly the reported findings suggested the proposed diagnosis.
2. It would describe in detail how and why the findings served to confirm or to rule out the proposed diagnosis.
3. It might suggest further findings that would help to confirm or rule out the diagnosis. Here again, it would give a detailed explanation as to why the findings were important.
4. Finally, the system might make certain further comments on the overall implication of the findings.

A key component of such a critique is the detailed discussion of the implications of the various diagnostic findings. By analyzing a physician's diagnosis in this fashion, a critiquing system could give the physician a great deal of information concerning the diagnostic process. In contrast, most diagnostic systems have focused their attention on diagnostic outcome: on producing a ranked list of possible diagnoses. In so doing, the rich structure of the diagnostic knowledge which went into obtaining the diagnoses is lost to the user. In fact, much of the most useful information, the detailed clinical explanations as to why certain findings are important, are generally not included in any current diagnostic systems. These descriptive details may be used by the domain expert in choosing a set of diagnostic rules to include in the system but are not part of the diagnostic rules themselves and are therefore not available to the user of the system.

In our preliminary exploration of critiquing in diagnostic radiology, it has become clear that a great deal of very detailed information about the diagnostic process is contained in radiologic textbooks and in the clinical literature. The radiologist has no hope of remembering all of this complex, detailed information, especially about diseases that he sees rarely.

The computer, on the other hand, has no trouble remembering this information. The critiquing approach allows the computer to bring this information to the radiologist selectively tailored to a particular case. As a result, a critiquing system can put the physician in touch with the clinical literature in a very natural way.

Our exploration of the critiquing of differential diagnosis is still in its early stages (P. L. Miller et al. 1985). It appears to hold promise, however, as a useful adjunct to the more traditional outcome-oriented approach to computer-assisted diagnosis.

10.2.2. Sophisticated Critiquing Capabilities

Another area for future research is in exploring further sophisticated critiquing capabilities. This research would involve basic computer science research, concentrating on fundamental design issues rather than on practical system development. Examples of this type of project are:

Critiquing treatment goals. The VQ-ATTENDING system currently infers a set of treatment goals that it uses to help critique management choices. As discussed in Chapters 5 and 9, the system might be extended to critique the physician's choice of treatment goals as well. This would extend critiquing beyond its current focus on the "tactics" of management to an analysis of more strategic considerations.

Critiquing interacting subplans. The systems described in this book assume that the various components of a plan are independent. As a result, the process of critiquing a part of a plan need not be based on a comprehensive analysis of the entire plan. On the other hand, several basic AI research projects in non-medical domains have explored the problems that occur when subplans interact (Sacerdoti 1977; Stefik 1981). If critiquing were extended to such a domain, it would presumably be necessary to coordinate different parts of the critique to reflect the fact that the various decisions are interdependent.

Critiquing based on fundamental biomedical principles. As discussed in Chapter 9, the systems described in this book base their analysis either on knowledge of established techniques or, in the case of the VQ-ATTENDING, on knowledge of treatment goals. In certain domains, however, knowledge of fundamental biomedical principles might be required to allow "de novo" construction of a critique based on a "deeper" understanding of the medical issues involved.

Critiquing based on knowledge of the clinical literature. The "deep knowledge" used by clinicians in making medical decisions, however, is frequently based on the clinical literature. As discussed in Chapter 9, it would be interesting to explore how such knowledge could be integrated into the critiquing process in a sophisticated fashion.

The projects outlined above involve basic computer science research since they confront complex intellectual problems. One of the best ways to explore such problems is by the prototype implementation of experimental systems. This indeed is how a great deal of computer science research is performed. The goal in building these prototypes is not immediate practical use (which involves a set of very different design issues), but to gain concrete feedback as to how the solution to a complex problem might be structured in a clear, conceptually elegant way.

10.2.3. Automated Knowledge Acquisition and Verification

Chapter 8 discussed the knowledge exerciser program that currently helps a domain expert interact with knowledge stored in the ESSENTIAL-ATTENDING format. Although this program is a useful tool, it is very simple. One limitation is that it knows nothing about the knowledge stored in the expressive frames it manipulates.

We are currently embarking on a project that will explore how underlying knowledge of the domain may allow the system itself to assist more actively, both in knowledge acquisition and knowledge verification. The ability to help a domain expert input his knowledge to an expert system is most important if such systems are ever to come into use on a large scale.

Our current project explores how a system that critiques radiologic workup could utilize information about the tests used (indications, contraindications, findings, etc.) together with a flowchart of the preferred workup sequence. This underlying knowledge would then guide the domain expert in filling out the various comments needed to critique the workup process. Using the flowchart and the other information supplied, the system itself could then anticipate many of the comments that might be required.

1. It could make sure that appropriate comments were included to deal with the various indications and contraindications of the various tests.
2. It could anticipate various ways that tests might be performed out of sequence and ask the domain expert for comments to make when this happened.
3. If different tests could produce conflicting findings, the system could make sure the domain expert had included appropriate comments to handle this situation.
4. The system might also suggest different possible sequences for presenting the material in the critique and might also suggest alternative paragraph structures for assembling certain parts of the analysis.

In this way, the system could use fundamental knowledge about the workup process, together with a flowchart of the problem at hand, to lead the domain expert by the hand in assembling a critiquing system. Radiologic workup may prove well suited to exploring these issues, since most areas are quite constrained and have a similar structure (the ordering of a sequence of tests). This project is still in the early stages of development.

10.2.4. Sophisticated Structuring of the Prose Critique

The critiquing systems described in this book all produce their prose critiques in a fairly regimented way. Although the material presented by a system may vary widely depending on the particular case, the overall structure and order of the presentation is quite rigid. The systems usually analyze and critique the components of the proposed plan in a predetermined sequence.

An interesting computational linguistics project might explore how a critiquing analysis could be structured differently depending on the particular content being discussed. This more sophisticated structure could better reflect such information as (1) key points, (2) key errors, (3) underlying treatment goals, (4) fundamental misconceptions, etc. This project would seek in some sense to structure the critique as closely as possible to the physician's own mental model of his plan.

10.2.5. Validation and Clinical Evaluation

While it is important to continue to explore underlying design issues, it is equally important to identify domains where practical critiquing systems can be implemented. Once such a system is built, it is then important to validate its knowledge, to ascertain that it is in some sense accurate, complete, and consistent. Later studies could evaluate the clinical efficacy of the system's advice. Sophisticated validation and evaluation of an expert medical system are complex research issues in their own right (P. L. Miller 1985b).

Validation. A major problem in validating an expert medical system is that there is seldom an absolute "gold standard" for comparison. Given a description of a particular case, different medical experts frequently come to different conclusions and may make different recommendations. As a result, the question of whether a system is "right" or "wrong" may be very fuzzy and difficult to determine in a rigorous way.

Clinical evaluation. A host of different parameters might be measured in attempting to evaluate an expert system's clinical performance. These issues range from quality of care, ease of use, ultimate effect on patient health, cost savings, etc. Since very few AIM systems have progressed to the operational stage, this whole area of research remains largely unexplored.

The task of validating and evaluating a critiquing system introduces further complexity, since a critiquing system reacts to a physician's plan, rather than recommending action itself. As a result, critiquing advice is an indirect reflection of the clinical decision being made, and therefore is one step away from the clinical domain. This makes formal evaluation more difficult, since it is even harder to determine an appropriate standard of comparison.

Indeed, much of the information imparted by a critiquing system is of a descriptive nature. A critique may include nuances and subtleties of meaning that may prove very hard to incorporate fully into a rigorous validation. In fact, if one thinks of a critiquing system, such as HT-ATTENDING, as an interactive paper, conventional forms of expert system validation may prove inappropriate.

On the one hand, one can certainly measure whether a critiquing system catches important errors and does not advocate alternatives that are clearly wrong. On the other hand, however, given (1) the subjectivity and variation in medicine, and (2) the descriptive character of a critiquing system's analysis, it may be especially difficult to determine rigorously whether the system's advice is right or wrong.

In fact, the ultimate utility of an "interactive paper" (once it has been demonstrated to contain no clear-cut errors) presumably rests on the subjective reaction of the user. This is certainly how a journal article or book is judged. Unfortunately, however, this reaction may be colored by how comfortable the user is with computer technology in the first place. In any case, it is clear that many interesting issues remain to be confronted in the validation and evaluation of critiquing systems.

10.2.6. Dissemination of Critiquing Systems for Practical Use

A final issues involves exploring different modes for disseminating a critiquing system for practical use. From the standpoint of technology, there are two main alternatives:

Network access. One alternative would be to implement the system on a single host computer, available to users via a nationwide computer network. This approach has several advantages. Only one copy of the system would exist, and periodic updating of the knowledge base would be greatly facilitated. Also, anyone with a "dumb" terminal (or personal computer) and a modem could access the system.

Software distribution. The other approach would be to distribute the system to users. This might be in the form of a disk to run on a personal computer, a program to run on a hospital computer system, etc. This alternative might prove cheaper if the system were frequently used. It has the disadvantages that (1) any change would have to be sent to all the users, and (2) interested users might have different, incompatible hardware.

Beyond these technological problems of dissemination, there are also legal problems as to who is responsible for any errors, or for a patient who merely suffers an adverse result. These issues all remain for the future.

Appendices

Appendix I: ESSENTIAL-ATTENDING
Implementation Overview

Appendix I discusses several practical issues involved in adapting E-ATTENDING to a different system environment. The various LISP programs that comprise E-ATTENDING are shown in Appendices II through V. The knowledge exerciser program is shown in Appendix VI.

Why LISP? E-ATTENDING is implemented in the LISP programming language (Winston and Horn 1981). For purposes of disseminating E-ATTENDING, we might have reprogrammed the system in a more conventional language such as PL/1 or PASCAL. We have elected not to do this, for several reasons:

1. E-ATTENDING is already written in LISP, which is a very well-designed language for symbolic programming.
2. E-ATTENDING takes advantage of a number of LISP features, including the ability to incorporate fragments of LISP code into data structures. It is awkward to implement this capability in many languages.
3. LISP is becoming increasingly available on smaller machines. Also, as the current trend of decreasing cost and increasing power in computer hardware continues, any computational inefficiencies of LISP will become increasingly unimportant.

RUTGERS-UCI LISP. E-ATTENDING is written in RUTGERS-UCI LISP, a dialect described in detail by Meehan (1979). In implementing E-ATTENDING, we have tried not to use esoteric features of this dialect. Nevertheless, some program modification is inevitable in adapting the system to another LISP dialect. These modifications should be purely syntactic and relatively few.

A Strategy for Implementing E-ATTENDING. A reasonable strategy for implementing E-ATTENDING is: (1) copy the LISP code in Appendices II through

V, (2) make any necessary syntactic changes to adapt it to a different LISP dialect, and then (3) debug the system using the simple example of Chapter 7.

Appendix II: The Production Rule Interpreter

The following routines coordinate the operation of E-ATTENDING's production rule interpreter. A more detailed overview of a similar interpreter can be seen in Chapter 18 of Winston and Horn (1981). As discussed in Chapter 7, the production rules operate on "facts," expressed as "triples" of the form:

(NAME ATTRIBUTE VALUE).

FACTS is a global variable used to store these "facts" and must be initialized to NIL.

DEFFACT asserts a few fact, e.g., (DEFFACT '(GOAL1 PRIORITY HIGH)).

GENRULE processes a rule. It evaluates the rule's IF-clause and, if appropriate, asserts the facts listed in the rule's THEN-clause.

SAME and TEST are functions that may be used in the IF-clause of a rule and are described in Chapter 7. (SAME and TEST are both defined as LISP "FEXPRs," which means that they receive their arguments unevaluated.)

```
(PROG ()
    (SETQ FACTS NIL)
    )

(DE DEFFACT (THEN)
    (PROG (NAME ATTRIBUTE VALUE OLD)
    (COND ((NULL THEN) (RETURN)))
    (COND ((OR (ATOM THEN) (NULL (CDR THEN)))
            (MSG T "error: " THEN " not a triple." -2) (RETURN)))
    (SETQ NAME (GETVAL (CAR THEN)))
    (SETQ ATTRIBUTE (GETVAL (CADR THEN)))
    (SETQ VALUE (COND ((NULL (CDDR THEN)) T) (T (CADDR THEN))))
    (SETQ OLD (GET1 FACTS NAME))
    (SETQ FACTS (PUT1 FACTS (PUT1 OLD VALUE ATTRIBUTE) NAME))
    ))

(DE GETVAL (A)
    (COND ((ATOM A) A)
        (T (EVAL A))))

(DE GENRULE (RULE)
    (COND ((EVAL (CAR (CDADDR RULE))) (MAPC 'DEFFACT (CDR (CADDDR
            RULE))))
        ((CDDDDR RULE) (MAPC 'DEFFACT (CDAR (CDDDDR RULE))))
        (T NIL)))
```

```
(DF SAME (IF)(APPLY# 'TEST (CONS 'EQUAL IF)))

(DF TEST (ARGS)
    (PROG (VALUE FUN)
    (SETQ VALUE (COND ((NULL (CDDDR ARGS)) T) (T (GETVAL (CADDDR
        AGRS)))))
    (SETQ FUN (CAR ARGS))
    (RETURN (APPLY# FUN (LIST (VAL (CADR ARGS)(CADDR ARGS))
        VALUE)))
    ))

(DE VAL (NAME ATTRIBUTE)
    (PROG ()
    (SETQ NAME (GETVAL NAME))
    (SETQ ATTRIBUTE (GETVAL ATTRIBUTE))
    (RETURN (GET1 (GET1 FACTS NAME) ATTRIBUTE)) ))
```

Appendix III: PROSENET

These routines coordinate the operation of PROSENET, E-ATTENDING's prose generator.

DEFPROSE converts the PROSENET ATNs to an internal format.
GENPROSE takes as its argument the name of an ATN state and coordinates the generation of prose starting at that state, invoking RUNOFF to print the prose.
RUNOFF actually prints the prose, performing the following functions:

1. Making sure each line length is appropriate
2. Translating punctuation (e.g., *comma, *period *para) appropriately
3. Capitalizing words at the start of sentences

```
PROG ()
    (SETQ MAX_CHAR 68)
    (SETQ NUM_CHAR 0)
    (SETQ PREVCHAR NIL)
    (SETQ CAPFLAG NIL)
    (SETQ CAPLETTERS '(
        q Q w W e E r R t T y Y u U i l o O p P a A s S d D f F g G
        h H j J k K l L z Z x X c C v V b B n N m M))
    )

(DE DEFPROSE (ATN) (MAPC 'SETSTATE ATN))

(DE SETSTATE (ST)
    (PROG (Z)
    (SETQ Z 1)
```

```
    (PUTPROP (CAR ST) (MAPCAR 'SETSTATE1 (CDR ST)) 'ARCS)
    ))
(DE SETSTATE1 (ARC)
    (PROG (A)
    (COND ((ATOM (CAR ARC)) (RETURN ARC))
          ((NULL (MEMQ (CAAR ARC) '(*OPTION *option)))(RETURN ARC)))
    (SETQ A (MAKNAM (APPEND (EXPLODE (CAR ST))
                            (APPEND '(_ T E M P _) (LIST Z)))))
    (SETQ Z (ADD1 Z))
    (INTERN A)
    (RETURN (CONS (CONS (CAAR ARC)(CONS A (CDAR ARC)))(CDR ARC)))
    ))
(DE GENPROSE (NODE)
    (PROG (ARCL ARC FLG T1 T2)
 STRT (SETQ ARCL (GET NODE 'ARCS))
  A (SETQ ARC (CAR ARCL))
  A1 (COND ((NULL (TESTQ ARC)) (GO B))
           ((TESTPOP ARC) (RETURN))
           ((TESTENDSEQ ARC) (GO F))
           ((TESTJUMP ARC) (GO C))
           ((ATOM (CAR ARC)) (RUNOFF (CAR ARC)))
           ((TESTFUNC ARC) (EVAL (CADAR ARC)))
           ((TESTPUSH ARC) (DOPUSH))
           ((TESTOPT ARC) (GO D))
           ((TESTSEQ ARC) (GO E))
           (T (RUNOFF (CAR ARC))))
    (GO C)
  E (SETQ T1 (CADAR ARC))
    (SETQ T2 (GET 'AAA T1))
    (COND ((NULL T2) (SETQ T2 (CDDAR ARC))))
    (SETQ ARC (CAR T2))
    (COND ((CDR T2) (PUTPROP 'AAA (CDR T2) T1)))
    (GO A1)
  F (SETQ T1 (CADAR ARC))
    (PUTPROP 'AAA NIL T1)
    (GO C)
  D (SETQ T1 (CADAR ARC))
    (SETQ T2 (GET 'AAA T1))
    (COND ((NULL T2) (SETQ T2 (CDDAR ARC))))
    (SETQ ARC (PICKOPT T2))
    (COND ((NULL (CDR T2)) (SETQ T2 (CDDAR (CAR ARCL)))))
    (PUTPROP 'AAA (DELOPT T2 ARC) T1)
    (GO A1)
```

```
    B (SETQ ARCL (CDR ARCL))
      (COND ((NULL ARCL)
                (MSG "ERROR: no more PROSENET arcs from state " NODE
             T)))
      (GO A)
    C (SETQ NODE (CADR ARC))
      (GO STRT)
      ))

(DE DOPUSH ()
    (PROG (A)
    (SETQ A (CADAR ARC))
    (GENPROSE (COND ((ATOM A) A)
                    (T (EVAL A))))
    ))

(DE TESTQ (ARC) (EVAL (CADDR ARC)))

(DE TESTSEQ (ARC) (MEMQ (CAAR ARC) '(*SEQUENCE *sequence)))

(DE  TESTENDSEQ  (ARC)  (MEMQ  (CAAR  ARC)  '(*ENDSEQUENCE
         *endsequence)))

(DE TESTPOP (ARC) (MEMQ (CAR ARC) '(*POP *pop)))

(DE TESTJUMP (ARC) (MEMQ (CAR ARC) '(*JUMP *jump)))

(DE TESTFUNC (ARC) (MEMQ (CAAR ARC) '(*FUNCTION *function)))

(DE TESTPUSH (ARC) (MEMQ (CAAR ARC) '(*PUSH *push)))

(DE TESTOPT (ARC) (MEMQ (CAAR ARC) '(*OPTION *option)))

(DE PICKOPT (ARCL) (EL (RNDN (LENGTH ARCL)) ARCL))

(DE DELOPT (L E)
         (COND ((NULL L) NIL)
               ((EQ (CAR L) E) CDR L))
               (T (CONS (CAR L) (DELOPT (CDR L) E)))))

(DE RUNOFF (TXT)
    (PROG ()
    (COND ((ATOM TXT)(SETQ TXT (LIST TXT))))
    A (COND ((NULL TXT)(RETURN))
        ((AND (NULL (ATOM (CAR TXT)))
            (MEMQ (CAAR TXT) '(*FUNCTION *function)))
                (EVAL(CADAR TXT)) (GO E))
        ((MEMQ (CAR TXT) '(*PARA *para)) (GO B))
        ((MEMQ (CAR TXT) '(*PERIOD *COMMA *COLON *OPENPAREN
                *CLOSEPAREN *APOST *APOST_S *QUOTE
```

```
                             *UNQUOTE *period *comma *colon *openparen
                             *closeparen *apost *apost_s *quote *unquote))
                             (GO C))
               ((MEMQ (CAR TXT) '(*NEWLINE *newline)) (GO F))
               ((GREATERP (PLUS NUM_CHAR (LENGTH (EXPLODE (CAR
                             TXT)))) MAX_CHAR)
                             (GO D)))
        (COND ((GREATERP NUM_CHAR O) (PRINC " ")))
        (COND (PREVCHAR (PRINC PREVCHAR) (SETQ PREVCHAR NIL)
                             (SETQ NUM_CHAR (ADD1 NUM_CHAR))))
        (PRINC (COND (CAPFLAG (CAPITALIZE (CAR TXT))) (T (CAR TXT))))
        (SETQ CAPFLAG NIL)
        (SETQ NUM_CHAR (PLUS 1 NUM_CHAR (LENGTH (EXPLODE (CAR
               TXT)))))
      E (SETQ TXT (CDR TXT))
        (GO A)
      B (TERPRI) (TERPRI) (PRINC "      ") (SETQ NUM_CHAR 5) (SETQ
               CAPFLAG T)
        (GO E)
      C (COND ((MEMQ (CAR TXT) '(*PERIOD *period))
                             (PRINC ". ")(SETQ CAPFLAG T)
                             (SETQ NUM_CHAR (ADD1 NUM_CHAR)))
               ((MEMQ (CAR TXT) '(*COMMA *comma)) (PRINC ","))
               ((MEMQ (CAR TXT) '(*COLON *colon)) (PRINC ":"))
               ((MEMQ (CAR TXT) '(*CLOSEPAREN *closeparen)) (PRINC ")"))
               ((MEMQ (CAR TXT) '(*UNQUOTE *unquote)) (PRINC "'"))
               ((MEMQ (CAR TXT) '(*APOST *apost)) (PRINC "'"))
               ((MEMQ (CAR TXT) '(*APOST_s *apost_s)) (PRINC "'s")
                             (SETQ NUM_CHAR (ADD1 NUM_CHAR))))
        (COND ((MEMQ (CAR TXT) '(*OPENPAREN *openparen)) (SETQ
                             PREVCHAR "("))
               ((MEMQ (CAR TXT) '(*QUOTE *quote)) (SETQ PREVCHAR "'")))
        (GO E)
      D (TERPRI)
        (SETQ NUM_CHAR O)
        (GO A)
      F (TERPRI)
        (SETQ NUM_CHAR O)
        (GO E)
        ))

(DE CAPITALIZE (WRD)
     (PROG (A B)
     (SETQ A (EXPLODE WRD))
     (SETQ B (GET1 CAPLETTERS (CAR A)))
```

```
(COND (B (RETURN (MAKNAM (CONS B (CDR A))))))
(RETURN WRD)))
```

Appendix IV: Routines That Manipulate Expressive Frames

The following routines manipulate E-ATTENDING's expressive frames.

DEFFRAME defines an expressive frame.

GENFRAME processes an expressive frame to generate prose.

GENCOMMENTS is the "standard" (builtin) routine used to coordinate the generation of prose comments of an expressive frame. It calls SELCOMMENTS to select those comments whose "IF conditions" are true, SORT-COMMENTS to sort the chosen comments according to their "SEQUENCE" values, and GENCOMMENT to invoke the prose generator on each comment in turn. Any of these four routines may be modified by the user, to allow more control of the prose-generating process.

```
(DE DEFFRAME (NAME COMM FCN)
    (PROG ()
    (PUTPROP NAME COMM 'COMMENTS)
    (PUTPROP NAME FCN 'FUNCTION) ))

(DE GENFRAME (FRM) ((GET FRM 'FUNCTION) (GET FRM 'COMMENTS)))

(DE GENCOMMENTS (COMMLIST)
    (MAPC 'GENCOMMENT (SORTCOMMENTS (SELCOMMENTS
        COMMLIST))))

DE SELCOMMENTS COMMLIST)
    (MAPCAN 'SEL1 COMMLIST))

(DE SEL1 (COMM)
    (COND ((EVAL (GET1 COMM 'IF)) (LIST COMM))
        (T NIL)))

(DE GENCOMMENT (COMM)
    (PROG (FCN FRM C)
    (SETQ FCN (GET1 COMM 'GENFUNCTION))
    (COND (FCN (EVAL FCN)))
    (SETQ C (GET1 COMM 'COMMENT))
    (COND ((NULL C)(GO A))
        ((ATOM C)(GENPROSE C))
        (T (RUNOFF C)))
  A (SETQ FRM (GET1 COMM 'FRAME))
    (COND (FRM (GENFRAME FRM)))
    ))
```

```
(DE SORTCOMMENTS (COMMLIST) (SORT COMMLIST 'SORTS1))

(DE SORTS1 (COMM1 COMM2) (LESSP (GETSEQUENCE COMM1)
          (GETSEQUENCE COMM2)))

(DE GETSEQUENCE (COMM)
    (PROG (A)
    (SETQ A (GET1 COMM 'SEQUENCE))
    (COND ((NUMBERP A) (RETURN A)))
    (RETURN O)))
```

Appendix V: Miscellaneous Utility Routines

The following are miscellaneous utility routines used by the system.

RNDN selects a random integer from 1 to N. It is used by the prose generator to
select an arbitrary OPTION arc. Since UCI LISP does not have a random number generator, the low order bits of the time of day (DTIME) are used.
GET1 and PUT1 are similar to the LISP primitives GET and PUTPROP except
that they operate on arbitrary lists of NAME VALUE pairs, e.g.:

$$(GET1 \text{ '(A 1 B 2) 'B)} = 2$$

$$(GET1 \text{ '(A 1 B 2) 'C)} = NIL$$

$$(PUT1 \text{ '(A 1 B 2) 5 'C)} = (A 1 B 2 C 5)$$

```
(DE RNDN (N)
    (PROG (TM TMP)
    (SETQ TM (DTIME))
    (SETQ TMP (CAR (DIVIDE TM 100)))
    (SETQ TMP (PLUS TM (TIMES - 100 TMP)))
    (RETURN (ADD1 (CAR (DIVIDE (TIMES TMP N) 100))))
    ))

(DE EL (N L) (CAR (NTH L N))))

(DE GET1 (L P)
    (COND ((NULL L) NIL)
          ((EQ (CAR L) P) (CADR L))
          (T (GET1 (CDDR L) P))))

(DE PUT1 (L V P)
    (COND ((NULL L) (COND ((NULL V) NIL)
                          (T (LIST P V))))
          ((EQ P (CAR L))(COND ((NULL V) (CDDR L))
                          (T (APPEND (LIST P V) (CDDR L)))))
          (T (CONS (CAR L) (CONS (CADR L) (PUT1 (CDDR L) V P))))))
```

Appendix VI: ESSENTIAL-ATTENDING's Knowledge Exerciser Program

This appendix shows ESSENTIAL-ATTENDING's knowledge exerciser program. The top-level program INSPECT takes as its input a list of names of expressive frames and operates as outlined in Chapter 8.

ASKFRAME displays the list to the user and returns the name of the selected frame.

ASKCOND compiles a list of all conditions that control the selected frame's prose output as described in Chapter 8, displays the list to the user, and then asserts the user's selected conditions as "facts."

```
(DE INSPECT (FRMLIST)
    (PROG (FRM)
    (MSG T" This is a relatively primitive augmentation of HT-ATTENDING that
    allows a domain expert to test the various components of the system's
    knowledge base, to see how the system reacts in different circumstances."
    T)
  A (MSG T"*** which of the following would you like to examine?")
    (SETQ FRM (ASKFRAME FRMLIST))
    (COND (NULL FRM) (RETURN)))
  B (MSG - 2 "*** please select a set of conditions - "T)
    (COND ((NULL (ASKCOND FRM)) (GO A)))
    (RUNOFF '* para)
    (GENFRAME FRM)
    (GO B)))

(DE ASKFRAME (L)
    (PROG (N ANS))
    (SETQ N 1)
    (AF1 L)
    (MSG "    type one (O to quit) **")
  A (SETQ ANS (READ))
    (COND ((MEMBER ANS ' (O O OK)) (RETURN))
          ((AND (NULL (NUMBERP ANS)) (ATOM ANS)) (RETURN ANS))
          ((OR (NULL (NUMBERP ANS)) (LESSP ANS 1) (GREATERP ANS
              (LENGTH L))) (MSG " please retype **") (GO A)))
    (RETURN (EL ANS D))))

(DE AF1 (D)
    (COND ((NULL L) NIL)
          (T (MSG N ". ") (AF2 (CAR L))
              (SETQ N (ADD1 N)) (AF1 (CDR L)) )))
```

```
(DE AF2 (A)
    (COND ((ATOM A) (MSG A T))
          ((NULL (EQUAL (CAR A) 'SAME)) (MSG A T))
          ((EQ (CADR A) 'HX) (POUT (OR (GET1 HXLST (CADDR A))
                (SMALL (CADDR A)))) (TERPRI))
          (T (PRINL (MAPCAR 'SMALL (CDR A))) (TERPRI))))

(DE POUT (X)
    (COND ((ATOM X)(PRINC X))
          (T (PRIN1 X))))

(DE ASKCOND (F)
    (PROG (CLIST A ANS)
    (MAPC 'AL1 (GET F 'COMMENTS))
    (SETQ N 1)
    (AF1 CLIST)
    (MSG "*** type one or more (O to quit) **")
  A (SETQ ANS (READL))
    (COND ((MEMBER ANS ' ((O) (O) (OK))) (RETURN))
          (MAPCAN (FUNCTION (LAMBDA (X)
                (COND ((OR (NULL (NUMBERP X)) (LESSP X 1)
                           (GREATERP X (LENGTH CLIST)))(LIST T))
                           (T NIL)))) ANS) (MSG " please retype **")
                           (GO A)))
    (SETQ FACTS NIL)
    (MAPCAR (FUNCTION (LAMBDA (X) DEFFACT (CDR (EL X CLIST))))) ANS)
    (RETURN T)
    ))

(DE AL1 (L) (PROG () (AL2 (GET1 L 'IF) (AL3 (GET1 L 'COMMENT))))

(DE AL2 (IF)
    (COND ((ATOM IF) (NIL)
          ((AND (EQUAL (CAR IF) 'SAME) (NOT (MEMBER IF CLIST)))
                (SETQ CLIST (APPEND CLIST (LIST IF))))
          (T (MAPC 'AL2 IF))))

(DE AL3 (C)
    (COND ((NULL C) NIL)
          ((EQUAL C T) NIL)
          ((NULL (ATOM C)) NIL)
          (T (MAPC 'AL4 (GET C 'ARCS)))))

(DE AL4 (ARC)
    (PROG ()
    (AL2 (EL 3 ARC))
    (AL3 (EL 2 ARC))))
```

References

Aikins, J.S., Kunz, J.C., Shortliffe, E.H., and Fallat, R.J.: PUFF: An experimental system for interpretation of pulmonary function data. Computers and Biomedical Res. 16:199–208, 1983.

Banks, G., and Weimer, B.: Symbolic coordinate anatomy for neurology (SCAN). Proceedings of the Seventh Symposium on Computer Applications in Medical Care, Washington, DC, 828–830, 1983.

Black, H.R.: Evaluation and treatment of the hypertensive patient. Primary Care 10:3–20, 1983.

Black, H.R., and Bursten, S.L.: A clinical scoring system for detection of patients with pheochromocytomas. Yale Journal of Biology and Medicine 57:259–272, 1984.

Blois, M.S.: Clinical judgement and computers. N. Eng. J. Med. 303:192–197, 1980.

Bravo, E.L.: Pheochromocytoma: current concepts in diagnosis, localization, and management. Primary Care 10:75–86, 1983.

Buchanan, B.G., and Shortliffe, E.H.: Rule-Based Expert Systems: The MYCIN Experiments of the Heuristic Programming Project. Reading: Addison-Wesley, 1984.

Civetta, J.M.: Goal-directed ventilatory support. In: Current Reviews in Respiratory Therapy. Current Reviews, Miami, Fla., 1983.

Clancey, W.J.: Details of the revised therapy algorithm. In: Buchanan, B.G., and Shortliffe, E.H., Eds.: Rule-Based Expert Systems. Reading: Addison-Wesley, 1984.

Clancey, W.J., and Letsinger, R.: NEOMYCIN: Reconfiguring a rule-based expert system for application to teaching. Proceedings of the Seventh International Joint Conference on Artificial Intelligence, Vancouver, 829–836, 1981.

Clancey, W.J., and Shortliffe, E.H. (Eds.): Readings in Medical Artificial Intelligence: The First Decade. Reading: Addison-Wesley, 1984.

Davis, R., and Lenat, D.B.: Knowledge Based Systems in Artificial Intelligence. New York: McGraw Hill, 1982.

Fagan, L.M., Kunz, J.C., Feigenbaum, E.A., and Osborn, J.J.: Representation of dynamic clinical knowledge: Measurement interpretation in the intensive care unit. Proceedings of the Sixth International Joint Conference on Artificial Intelligence, Tokyo, 260–262, 1979.

Gorry, G.A., Silverman, H., and Pauker, S.G.: Capturing clinical expertise: A computer program that considers clinical responses to digitalis. Am. J. Med. 64:452–460, 1978.

Harrison, M.J., and Johnson, F.: Codifications of anesthetic information for computer processing. J. Biomed. Engr. 3:196–199, 1981.

Kastner, J., Weiss, S., and Kulikowski, C.: Treatment selection and explanation in expert medical consultation: application to a model of ocular herpes simplex. Proceedings of MEDCOMP-82, Philadelphia, 420–427, 1982.

Kingsland, L.C., Gaston, L.W., Vanker, A.D., and Lindberg, D.A.B.: A knowledge-based consultant system for problems in human hemostasis. Proceedings of the American Medical Informatics Association Congress-82, San Francisco, 325–329, 1982.

Kulikowski, C.A.: Artificial intelligence methods and systems for medical consultation. IEEE Trans. PAMI PAMI-2:464–476, 1980.

Langlotz, C.P., and Shortliffe, E.H.: Adapting a consultation system to critique user plans. International Journal of Man-Machine Studies 19:479–496, 1983.

Lindberg, D.A.B., Sharp, G.C., Kingsland, L.C., Weiss, S.M., Hayes, S.P., Ueno, H., and Hazelwood, S.E.: Computer based rheumatology consultant. Proceedings of MEDINFO-80, Philadelphia, 1311–1315, 1980.

Lindsay, R.K., Buchanan, B.G., Feigenbaum, E.A., and Lederberg, J.: Applications of Artificial Intelligence for Organic Chemistry: The DENDRAL Project. New York: McGraw-Hill, 1980.

London, R., and Clancey, W.J.: Plan recognition strategies in student modelling: prediction and description. Proceedings of the American Association for Artificial Intelligence, Pittsburgh, 335–338, 1982.

Long, W.J., and Russ, T.A.: A control structure for time dependent reasoning. Proceedings of the International Joint Conference on Artificial Intelligence, Karlesruhe, West Germany, 1983.

Long, W.J., Naimi, S., and Criscitiello, M.G.: A knowledge representation for reasoning about the management of heart failure. Proceedings of the IEEE Conference on Computers in Cardiology, 373–375, 1982.

Long, W.J., Russ, T.A., and Locke, W.B.: Reasoning from multiple information sources in arrythmia management. Proceedings of IEEE-83: Frontiers of Engineering and Computing in Health Care, 1983.

Manger, W.M., and Gifford, R.W.: Pheochromocytoma. New York: Springer-Verlag, 1977.

Materson, B.J., Oster, J.R., Michael, U.F., Bolton, S.M., Burton, Z.C., Stambaugh, J.E., and Morledge, J.: Dose response to chlorthalidone in patients with mild hypertension. Efficacy of a lower dose. Clin. Pharmacol. Ther. 24:192–198, 1978.

Meehan, J.R.: The New UCI LISP Manual. Hillsdale, NJ: Lawrence Erlbaum, 1979.

Miller, M.L., and Goldstein, I.: Structured planning and debugging. Proceedings of the Fifth International Joint Conference on Artificial Intelligence, Boston, 773–779, 1977.

Miller, P.L.: Critiquing anesthetic management: the "ATTENDING" computer system. Anesthesiology 53:362–369, 1983a.

Miller, P.L.: ATTENDING: Critiquing a physician's management plan. IEEE Trans. PAMI PAMI-5:449–461, 1983b.

Miller, P.L.: A Critiquing Approach to Expert Computer Advice: ATTENDING. London/Boston: Pitman, 1984.

Miller, P.L.: Goal-directed critiquing by computer: ventilator management. Computers and Biomedical Res. 18:422–438, 1985a.

Miller, P.L.: Issues in the evaluation of artificial intelligence systems in medicine. Proceedings of the Ninth Symposium on Computer Applications in Medical Care. Baltimore, 281–286, 1985b.

Miller, P.L.: Building an expert critiquing system: ESSENTIAL-ATTENDING. Meth. Inf. Med., 1985c (in press).

Miller, P.L., and Black, H.R.: Medical plan-analysis by computer: Critiquing the pharmacologic management of essential hypertension. Computers and Biomedical Res. 17:38–54, 1984.

Miller, P.L., Blumenfrucht, S.J., and Black, H.R.: An expert system which critiques patient workup: modeling conflicting expertise. Computers and Biomedical Res. 17:554–569, 1984.

Miller, P.L., Shaw, C., Rose, J.R., and Swett, H.A.: Critiquing the process of radiologic differential diagnosis. Proceedings of the Ninth Symposium on Computer Applications in Medical Care. Baltimore, 182–185, 1985.

Miller, R.A., Pople, H.E., and Meyers, J.D.: Internist-I, an experimental computer-based diagnostic consultant for general internal medicine. N. Eng. J. Med. 307:468–476, 1982.

Patil, R.S., Szolovits, P., and Schwartz, W.B.: Causal understanding of patient illness in medical diagnosis. Proceedings of the Seventh International Joint Conference on Artificial Intelligence, Vancouver, 893–899, 1981.

Pauker, S.G., Gorry, G.A., Kassirer, J.P., and Schwartz, W.B.: Towards the simulation of clinical cognition: taking a present illness by computer. Am. J. Med. 60:981–996, 1976.

Politakis, P., and Weiss, S.M.: A system for empirical experimentation with expert knowledge. In: Clancey, W.J., and Shortliffe, E.H. (Eds.): Readings in Medical Artificial Intelligence: The First Decade. Reading: Addison-Wesley, 1984.

Pople, H.E.: Heuristic methods for imposing structure on ill-structured problems: The structuring of medical diagnostics in artificial intelligence in medicine. In: Szolovits, P. (Ed.): Artificial Intelligence in Medicine. Boulder: Westview Press, 1982.

Reggia, J.A., and Perricone, B.T.: Knowledge-based decision support systems: Development through high-level languages. Proceedings of the Twentieth Annual Technical Symposium of the Association for Computing Machinery. Washington, D.C., 72–82, 1981.

Sacerdoti, E.D.: A Structure for Plans and Behavior. New York: Elsevier, 1977.

Scott, A.C., Clancey, W.J., Davis, R., and Shortliffe, E.H.: Methods of generating explanations. In: Buchanan, B.G., and Shortliffe, E.H. (Eds.): Rule-Based Expert Systems. Reading: Addison-Wesley, 1984.

Shapiro, B.A., Harrison, R.A., and Walton, J.R.: Clinical Application of Blood Gases. Chicago: Year Book Medical Publishers, 1977.

Shortliffe, E.H.: Computer-Based Medical Consultations: MYCIN. New York: Elsevier, 1976.

Shortliffe, E.H., Buchanan, B.G., and Feigenbaum, E.A.: Knowledge engineering for medical decision making. A review of computer-based clinical decision aids. Proc. IEEE 67:1207–1224, 1979.

Shortliffe, E.H., Scott, A.C., Bischoff, M.B., Campbell, A.B., Van Melle, W., and Jacobs, C.D.: An expert system for oncology protocol management. Proceedings of the Seventh International Joint Conference on Artificial Intelligence, Vancouver, 876–881, 1981.

Stefik, M.: Planning with constraints (MOLGEN: part 1). Planning and meta-planning (MOLGEN: part 2). Artificial Intell. 16:111–170, 1981.

Suwa, M., Scott, A.C., and Shortliffe, E.H.: Completeness and consistency in a rule-based system. In: Buchanan, B.G., and Shortliffe, E.H. (Eds.): Rule-Based Expert Systems. Reading: Addison-Wesley, 1984.

Swartout, W.R.: Explaining and justifying expert consultation programs. Proceedings of the Seventh International Joint Conference on Artificial Intelligence, Vancouver, 815-822, 1981.

Szolovits, P. (Ed.): Artificial Intelligence in Medicine. Boulder: Westview Press, 1982.

Teach, R.L., and Shortliffe, E.H.: An analysis of physician attitudes regarding computer-based clinical consultation systems. Computers and Biomedical Res. 14:542-558, 1981.

The 1980 Report of the Joint National Committee on Detection, Evaluation, and Treatment of High Blood Pressure. Arch. Intern. Med. 140:1280-1285, 1980.

van Melle, W.: A domain-independent production-rule system for consultation programs. Proceedings of the Sixth International Joint Conference on Artificial Intelligence, Tokyo, 923-925, 1979.

Weinstein, M.C., Fineberg, H.V., Elstein, A.S., Frazier, H.S., Neuhauser, D., Neutra, R.R., and McNeil, B.J.: Clinical Decision Analysis. Philadelphia: Saunders, 1980.

Weiss, S., and Kulikowski, C.: EXPERT: A system for developing consultation models. Proceedings of the Sixth International Joint Conference on Artificial Intelligence, Tokyo, 942-950, 1979.

Weiss, S., Kulikowski, C., Amarel, S., and Safir, A.: A model-based method for computer-aided medical decision-making. Artificial Intell. 11:145-172, 1978.

Weiss, S., Kulikowski, C., and Galen, R.: Developing microprocessor based expert models for instrument interpretation. Proceedings of the Seventh International Joint Conference on Artificial Intelligence, Vancouver, 853-855, 1981.

Winston, P.H., and Horn, B.K.P.: LISP. Reading: Addison-Wesley, 1981.

Woods, W.A.: Transition network grammars for natural language analysis. CACM 13:591-606, 1970.

Index

Computers and Medicine

Bruce I. Blum, *Editor*

Computers and Medicine